Edmund Knecht, Rudolf Benedikt

The Chemistry of the Coal-Tar Colours

Second Edition

Edmund Knecht, Rudolf Benedikt

The Chemistry of the Coal-Tar Colours
Second Edition

ISBN/EAN: 9783744790451

Printed in Europe, USA, Canada, Australia, Japan

Cover: Foto ©berggeist007 / pixelio.de

More available books at **www.hansebooks.com**

TECHNOLOGICAL HANDBOOKS.

THE CHEMISTRY

OF THE

COAL-TAR COLOURS.

TRANSLATED FROM THE GERMAN OF

DR. R. BENEDIKT,

AND EDITED, WITH ADDITIONS, BY

E. KNECHT, Ph.D., F.I.C.,

HEAD MASTER OF THE CHEMISTRY AND DYEING DEPARTMENT OF THE TECHNICAL
COLLEGE, BRADFORD; EDITOR OF THE "JOURNAL OF THE SOCIETY
OF DYERS AND COLOURISTS."

SECOND EDITION, REVISED AND ENLARGED.

LONDON: GEORGE BELL AND SONS, YORK STREET,
COVENT GARDEN.
1889.

BUTLER & TANNER,
THE SELWOOD PRINTING WORKS,
FROME, AND LONDON.

PREFACE.

ALTHOUGH England may rightly be called the birthplace of the coal-tar-colour industry, it is a remarkable fact that the English literature on the subject is very scanty, and that which does exist is now almost obsolete, owing to the rapid strides which have been made during the last ten years in the manufacture of coal-tar colours.

Numerous elaborate works are published from year to year on this important subject both in France and in Germany, but more especially in the latter country, by means of which manufacturers and students are continuously kept informed of the latest methods and theories. The want of a text-book of this kind containing a concise account of all the more important coal-tar colours has long been felt by the author, and has prompted him to undertake the present translation of Dr. Benedikt's excellent little work, "Die Künstlichen Farbstoffe," which has met with such success in all parts of the Continent where German is spoken. The author may therefore not be considered too sanguine in hoping that the present volume will meet with some success in this country, not only as a stepping-stone for those who wish to make the subject a special study, but also among students of chemistry in general. It will also be found useful to those engaged in the dyeing and printing of textile fabrics.

The manufacture of coal-tar colours is an industry which cannot possibly thrive by the old " rule of thumb " way of doing business. It is a well-known fact that those who succeed best in this industry now-a-days spare no means to bring the highest scientific training to bear on the subject, while those who are " penny wise and pound foolish " endeavour to save the expense of adequate scientific advice, and thus ultimately find themselves completely surpassed by more enlightened rivals.

E. K.

BRADFORD,
January, 1886.

PREFACE TO THE SECOND EDITION.

—

THE large number of new coal-tar colours which has come into the market since the publication of the first edition has necessitated a considerable increase in the size of the present volume. The author has also considered it advisable to rearrange the matter so as to bring the whole of the raw products under one heading, and to make that chapter as complete as the conciseness and scope of the work would allow.

During the same period the scientific development of the subject has gone rapidly ahead, and the constitutional formulæ of many products of which we then knew little more than the empirical formulæ have been definitely proved.

With the exception of these alterations, the general arrangement of subjects does not differ materially from that of the first edition.

At the suggestion of my friend, Mr. Christopher Rawson, the symbolic rendering of the " azo " group, which is generally represented as $-N=N-$, has been altered throughout the work to $-N:N-$, because the two bonds joining the nitrogen atoms might sometimes be misleading to beginners on account of the similarity they have to the sign " equals."

The author acknowledges his indebtedness to Drs. Schulz and Julius, for permission to make use of their excellent " Tabellarische Uebersicht der Künstlichen Farbstoffe " in compiling the tables of the azo-dyes on pp. 262–269.

E. K.

BRADFORD,
    May, 1889.

CONTENTS.

	PAGE
INTRODUCTION	1
THE OPTICAL PROPERTIES OF COLOURING MATTERS	5
GENERAL CHEMICAL PROPERTIES OF THE COLOURING MATTERS	27
METHODS OF DISSOLVING THE COLOURING MATTERS	35
DYEING WITH COAL-TAR COLOURS	39
THE TESTING OF COLOURING MATTERS	60
COAL-TAR AND THE RAW PRODUCTS USED IN THE MANUFACTURE OF THE COAL-TAR COLOURS	69
THE COAL-TAR COLOURS	139
I. ANILINE DYES	140
(a) THE ROSANILINE GROUP	140
(b) INDULINES AND SAFRANINES	186
(c) OXAZINES	196
(d) ANILINE-BLACK AND NAPHTHAMEÏN	201
(e) COLOURING MATTERS CONTAINING SULPHUR (THIO-NINES)	208
II. PHENOL DYE-STUFFS	212
(a) NITRO-BODIES	213
(b) COLOURING MATTERS PRODUCED BY THE ACTION OF NITROUS ACID ON PHENOLS	224
(c) ROSOLIC ACIDS	227
(d) PHTHALEÏNS	230
III. THE AZO DYES	247
(a) AMIDOAZO DYES	251
(b) AMIDOAZOSULPHONIC ACIDS	255
(c) OXYAZO DYES	259
DERIVATIVES OF QUINOLINE AND ACRIDINE	283
IV. ARTIFICIAL INDIGO	286
V. THE ANTHRACENE COLOURING MATTERS	291
INDEX	327

ERRATA.

Page 43, 9th and 10th lines from top, *read* " basic constituent or constituents," *instead of* " basic constituent or constituent."

Page 42, 9th line from bottom, *read* " It contains besides carbon," *instead of* " It contains, besides carbon."

Page 80, lines 4 and 7 from top, *read* " substitution derivatives," *instead of* " substitutive derivatives."

Page 282, line 15 from bottom, *read* " Primuline," *instead of* " Primuine."

Page 304, line 7 from top, *read* " Liechti," *instead of* " Lie-chti."

THE CHEMISTRY

OF THE

COAL-TAR COLOURS.

INTRODUCTION.

UNTIL the middle of this century the dyeing industry
was dependent upon those colouring matters which are
either found as such in the vegetable and animal kingdom,
or which are produced from some of the constituents of
the latter by very simple chemical processes. This whole
group of vegetable and animal colouring matters embraces
all the so-called *natural* colouring matters, while those
which the chemistry of modern times has evolved from
organic bodies possessing a comparatively simple composition, by operations which cause a total change of the
raw material, are generally designated as *artificial* colouring matters.

In the manufacture of these artificial colouring matters
only very few of the many organic substances employed
are obtained from the vegetable kingdom (*e.g.*, tannin,
which after its conversion into pyrogallol is used for
the preparation of cœruleïn). The greater part of the
materials which serve to furnish the artificial colouring
matters is obtained from coal-tar, a bye-product of the
manufacture of coal-gas.

It is owing to this fact that the history of the manufacture of artificial colouring matters, or "coal-tar colours,"

1   B

is to a great extent intimately connected with the history of the manufacture of coal-gas, and there is no doubt that the general introduction of coal-gas for illuminating purposes within the first half of the present century has made the manufacture of coal-tar colours possible. Nevertheless, from the 1st of April in 1814, when the parish of St. Margaret in Westminster was first illuminated by coalgas, a period of no less than forty-two years elapsed before the manufacture of the first aniline dye, mauveïne, was taken in hand.

During this long period the constituents of coal-tar were scientifically investigated, and thus a basis was formed on which the subsequent development of the coaltar colour industry rested.

Great difficulties were encountered in the study of coaltar, for sixty years ago organic chemistry was only in its childhood, and only with the gradual development of this science to its present position has our knowledge of the constituents of coal-tar become perfected. On the other hand, the chemistry of to-day has been furthered to a great extent by a thorough and incessant study of this bye-product.

But even at the present time our knowledge of the chemistry of coal-tar is by no means complete. We know that it consists of a mixture of a large number of compounds, about 60 of which have been obtained in the pure state; but we nevertheless suppose that it contains other compounds, which have hitherto not been isolated.

A short *résumé* of the dates of discovery of the most important constituents of coal-tar is given in the following:—

Naphthalene was first discovered in tar in 1820 by *Garden, anthracene* in 1832 by *Dumas*, and *phenol* in 1834 by *Mitscherlich*. *Benzene* was discovered in 1825 by *Faraday*, but its presence in coal-tar was only recognised in 1845 by *A. W. Hofmann*. *Toluene* was discovered in 1837 by *Pelletier* and *Walter*, and in 1848 *Mansfield* showed that it was contained in coal-tar.

Aniline was first discovered in 1826 by *Unverdorben* in the products of the dry distillation of indigo, and in 1834 *Runge* proved it to be a constituent of coal-tar. The latter contains it, however, in such small quantities that its isolation on a large scale would not pay. The production of aniline as a commercial product only became practicable when *Zinin* showed in 1842 that it could be produced by the reduction of *nitrobenzene*, a substance discovered in 1834 by *Mitscherlich*. *Béchamp* greatly improved this process in 1854 by the use of a mixture of iron and acetic acid as reducing agents. Within the last few years the method has been further improved by the employment of hydrochloric acid instead of acetic acid.

Runge first noticed in 1834 that aniline, when brought in contact with chloride of lime, gave brilliant colours; but it was not until 1856 that *Perkin* prepared *mauveïne*, the first aniline dye, on a large scale.

In 1858 *A. W. Hofmann* published a work on the action of carbon tetrachloride on aniline, by which reaction he obtained aniline-red. It was in 1859 that *Verguin* first manufactured aniline-red (magenta) in quantity.

During the following five years, *violet*, *blue*, and *green* colouring matters were invented and manufactured.

Aniline-black was discovered in 1863 by *Lightfoot*. *Graebe* and *Liebermann* effected in 1868 the synthesis of *alizarin*, the most valuable colouring principle of madder, a discovery which had the greatest influence on the whole colour industry.

The first *Eosin dye* was prepared in 1874 by *Baeyer*, while in later years a large number of important dye-stuff, such as the *azo dyes, methylene-blue, malachite-green*, etc., have been prepared. In the year 1880, Baeyer was so far advanced in his experiments on the preparation of artificial *indigo*, that the "Badische Anilin und Soda-fabrik" could venture to send into the market nitrophenyl-propiolic acid, a product by means of which indigo can be produced on the fibre.

The discovery of the *benzidine dyes*, of which *Congo red* and *Chrysamin* were the first to appear in the market. may be regarded as the most important discovery of modern times as far as cotton dyeing is concerned.

In going through the large number of artificial colouring matters which have been brought into the market since 1858, it will be seen that those products which are distinguished by superior brilliancy and fastness have soon taken the place of other colouring matters; to such an extent, indeed, that the manufacture of many dyes, which at one time flourished, has either had to be relinquished altogether, or has at least been considerably reduced. And although it may seem to an outsider, while looking through a collection of our modern silk and satin materials, that the dazzling and pure shades obtained cannot be surpassed, those engaged in the manufacture of these dyes are nevertheless well aware that even to-day they have not yet arrived at their ultimate aim. Of all the artificial dye-stuffs at present in use there are perhaps only a few, especially those which are distinguished by their superior fastness, that will maintain a permanent position in dyeing.

A popular prejudice still exists against the so-called " Aniline Dyes " as being far behind the animal and vegetable dye-stuffs with respect to fastness. But it is just those loose artificial colouring matters which are so soon replaced by other faster ones, and we now possess a considerable number of coal-tar colours which are just as fast, and often faster, than the natural ones.

PART I.

THE OPTICAL PROPERTIES OF COLOURING MATTERS.

THE immense diversity of objects which surround us appear to our eye in an infinite variety of colours and shades. But the light which they reflect is, as will be shown further on, very seldom perfectly uniform, and is in most cases composed of a greater or lesser number of rays of light differently coloured. The study of optics affords us the means of finding out these different elements, *i.e.*, of *analysing* the effect produced on our retina by any given colour.

The colour chemist can make use of the results obtained in this manner for the solution of many important questions, by the ordinary study of which he could gain little or no information.

The following will serve to illustrate a few questions of this kind :—

" What shades can be produced with a new colouring matter ? "

" Can these shades be produced with other colouring matters, already known ; or, cannot the shades obtained with the new colouring matter be obtained just as well with a mixture of two or more colouring matters, already known ? "

" Do these shades possess the greatest purity and brilliancy that can be expected of a colouring matter, or may the colouring matter be surpassed by another of its kind, perhaps one that has still to be discovered ? "

Primary colours.—However great the diversity of shades which we perceive, the number of primary impressions produced on the retina is very limited. These impressions are those of *white* and *black*, and of *red, yellow, green* and *blue.* The four last-named colours are termed *primary colours*, for no human eye has yet been able to detect in them two different colours, while all other colours contain two or more primary colours.

By mixing white, black, and the primary colours in different proportions, every impression is produced which the eye is capable of receiving. The number of impressions which can be produced by the mixture of these colours is, however, diminished by the fact that the primary colours will not produce an impression in every combination of two. Thus the impression of red can be mixed with blue or yellow, but not with green; for although we know a bluish-red and a yellowish-red, we do not know a reddish-green. If red and green rays, or blue and yellow rays, enter our eye simultaneously, we have the impression of white. When two colours unite to form white, they are called *complementary colours.* Green would therefore be the complementary colour of red, blue that of yellow.

Since, therefore, it is not possible for us to distinguish in a colour red and green, or blue and yellow, the following colours may be produced without the assistance of white and black :—

Red—reddish-yellow—yellow—yellowish-green—green —greenish-blue—blue—bluish-red.

These colours are called *saturated* as long as they do not contain any *white.* The saturation is not diminished by an addition of *black*, but the *intensity* of the colours is diminished. The different gradations are called *shades.* Thus it is possible to produce any number of shades between blue and red, of which the first will contain a large proportion of blue with very little red, while the last contain a large proportion of red and very little blue. The shades produced in this manner are known as bluish-violet, violet, reddish-violet, purple, and crimson.

All gradations or shades of grey lie between white and black. All the primary colours and their shades or mixtures can be mixed with white or any shade of the grey series, and thus an immense number of new colours would be produced which would represent the shades obtained with primary colours and shades.

Thus, by the construction of our eye and of our brain, we cannot distinguish, even in the most complicated colour, more than four elements; viz., two primary colours, white and black.

The marks or brands generally used in practice for distinguishing the shades of a colour may be mentioned here. For this purpose certain initials are placed after the name of the colouring matter, viz.: R for red (roth, rouge), O for orange, G or J for yellow (gelb, jaune), B for blue (blau, bleu), and V for violet. Thus we have a Scarlet G and a Scarlet R, the redder shades being denoted by the terms Scarlet RR, Scarlet RRRR or 4R, etc. Ordinary aniline-blue possesses a red shade, and is known as Aniline blue R, while the finest qualities are known as Aniline-blue 6B, etc.

Decomposition of light by the prism.

White light contains a number of different coloured rays, which are distinguished from each other by the length of their waves and number of oscillations. By means of the prism it is possible to isolate these rays, and to arrange them systematically according to their wave-lengths.

This can be demonstrated by the following simple experiment:—Direct or reflected sunlight is allowed to pass through a narrow slit A into a dark room. By means of the lense L, the light passing through the slit is made to converge so as to form a distinct image on the screen SS.

If now the rays are intercepted by the prism P, the white image at B will disappear, and a much broader

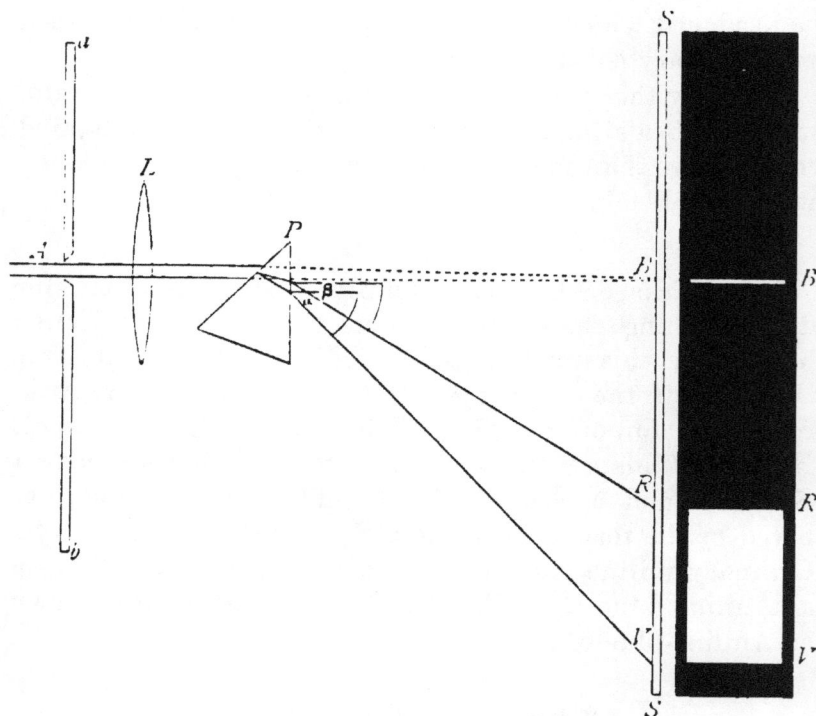

Fig. 1.

image RV, consisting of coloured bands, will be formed on another part of the screen. White light contains, therefore, rays which are refracted from B to V (by the angle a), and those which are refracted least, to R, along with a number of intermediate ones. The rays which are refracted most produce in the normal eye the impression of violet, and then the other rays follow, passing through bluish-violet to blue, bluish-green to green, greenish-yellow, yellow and orange to red. This coloured image is called a *spectrum*. The spectrum of white light contains all the pure colours, with the exception of those lying between the red and the violet; viz., crimson, purple, etc.

Every point of the spectrum consists of *homogeneous* light; *i.e.*, light which cannot be further decomposed. This can easily be proved by a slight modification of the

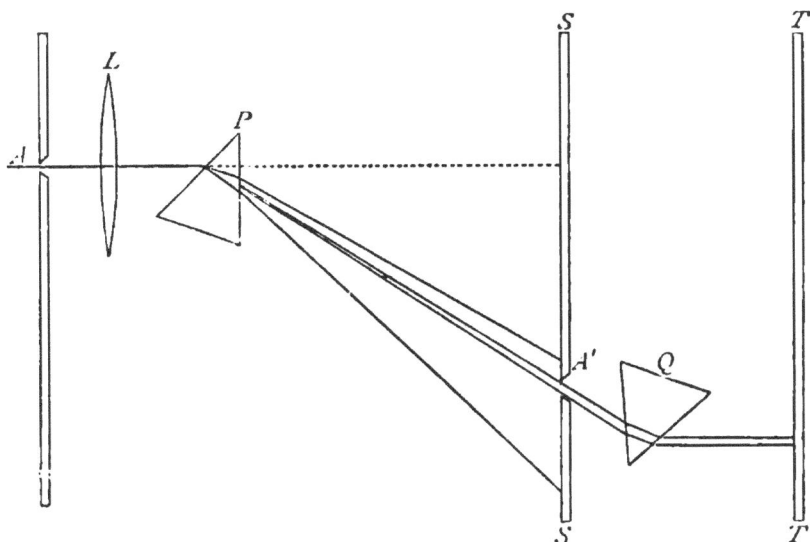

Fig. 2.

previous experiment. A narrow slit A' is made, parallel to A in the screen $S\,S$, which can be adjusted so as to allow any portion of the spectrum to pass through. The light passing through the slit A is caused to pass through a second prism Q, and is cast on to a second screen $T\,T$. It will then be seen that the orange rays, for instance, cannot be resolved into red and yellow, as might have been expected; the new image remains orange.

The impression of orange can therefore be produced in

our eye by homogeneous light of a certain refrangibility.
But when a mixture of red and yellow rays (that is, of
rays of greater and less refrangibility respectively than
the orange rays) enters our eye, we have again the impression of orange.

The naked eye is not capable of discerning whether an
orange is homogeneous or mixed, and it is furthermore,
as already stated above, not able to analyse colours or to

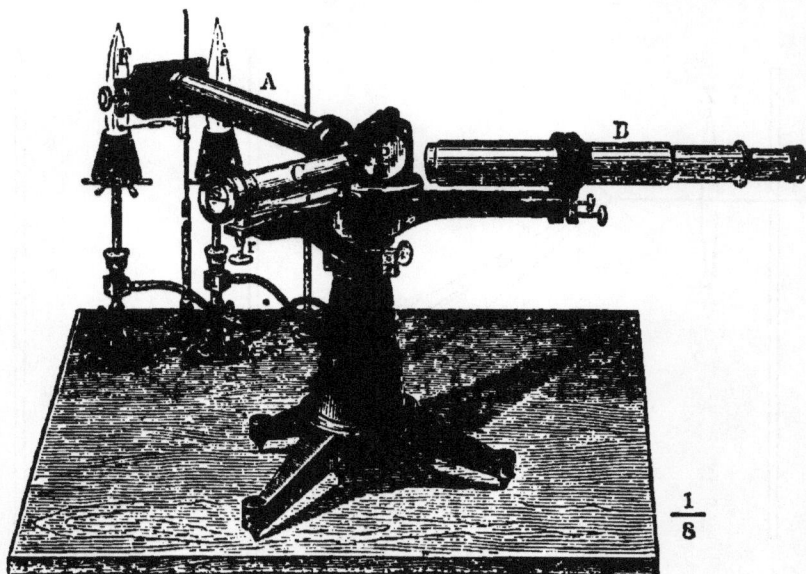

FIG. 3.

recognise the different coloured rays which produce a
certain impression of colour. It is invariably necessary
to make use of the prism or of some similar arrangement
in order to obtain an exact idea of the different compo-
nents of a colour. A convenient form of apparatus for
this purpose is the *spectroscope*, in which the spectrum
produced by the prism is not thrown on to a screen, but
is seen directly.

One of the best forms of spectroscope is the one given

in Fig. 3. The light to be analysed enters the tube A through a slit, the width of which can be regulated by means of a screw. At the end of this tube there is a lens, which is constructed so that the slit lies in the focus, and the rays are thus caused to pass, parallel to each other, into the prism P. The tube B is an astronomical telescope, through which an image of the spectrum can be seen. It can be turned around a vertical axis, and in this way every part of the spectrum can be brought into the middle of the plane of vision. The tube C also contains, at the end adjoining the prism, a lens, in the focus of which is placed a negative photographic image of a millimeter scale. This tube is placed so that its axis forms with the front face of the prism an angle equal to the one formed by the axis of the tube B, so that an eye looking through B sees simultaneously the illuminated scale and the spectrum, an arrangement which is of great advantage in the description and the comparison of the spectra.

Another form of spectroscope is known as the *spectroscope with direct vision*. This form consists of a single tube, containing, besides the slit and the lens, two or more prisms composed of Crown glass and Flint glass alternately.

The Spectra.

All solid or liquid bodies when heated to a white heat give a continuous spectrum; *i.e.*, one which is not interrupted by any dark lines or bands.

The rays which are emitted from the white-hot substance of the sun have to pass, before reaching us, through the sun's atmosphere, and in consequence of this the sun's spectrum is not a continuous one, but is traversed by a large number of fine lines, known as Fraunhofer's lines. Fig. 4 shows, according to Rood, the arrangement of the colours in the sun's spectrum and the relative positions of the most important of Fraunhofer's lines,

Incandescent gases or vapours, on the other hand, give *discontinuous* spectra, *i.e.*, spectra in which the number of rays of light is limited, and they appear in the spectroscope only as lines of the breadth of the slit. These spectra are called *line-spectra*, and every chemical element possesses in the incandescent gaseous condition its own characteristic lines by means of which it can be recognised.

The simplest way of observing these spectra is to place the non-luminous flame of a Bunsen's burner in front of the slit of the spectroscope, and to introduce into it on the end of a platinum wire a volatile salt of the element to be examined.

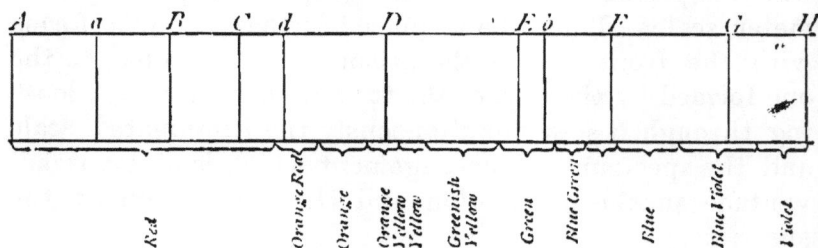

FIG. 4.

These spectra are produced by *luminous* bodies. The colouring matters, their solutions and the substances dyed with them, are not luminous in themselves, but they possess the property of converting white light which strikes or traverses them into coloured light. The explanation of this is, that they reflect or are traversed by only a portion of the rays contained in white light, the rest being hidden or destroyed by absorption. If, for instance, a glass vessel containing a dilute solution of magenta is placed between a luminous flame, which gives a continuous spectrum, and the slit of a spectroscope, it will be seen on analysing the transmitted light that an essential change has taken place. Only the red rays of the spectrum pass through unchanged; the greenish-yellow ones are extinguished, and blue and violet are considerably weakened. The result is, therefore, a discontinuous *absorption-spectrum*, which consists of two parts, one containing a large number of red rays and the other the bluish-violet rays considerably diminished in intensity, while the two portions are separated from each other by a black interspace.

The light which strikes our eye from coloured surfaces is not simply reflected light; part of it has traversed the uppermost layers of the coloured body, and being reflected in the interior has then entered our eye, partially deprived of its coloured rays. This coloured light is invariably mixed with a certain proportion of white light, reflected from the actual surface of the body before having entered the uppermost layer.

If, therefore, dyed fabrics are to be examined by spectrum analysis, the same phenomena are in general observed as the solutions of the corresponding colouring matters show when in a state of solution. In both cases an *absorption-spectrum* is obtained; but the one given by the solution is the purer, because it does not contain the admixture of white light which is reflected from the surface of the coloured objects.

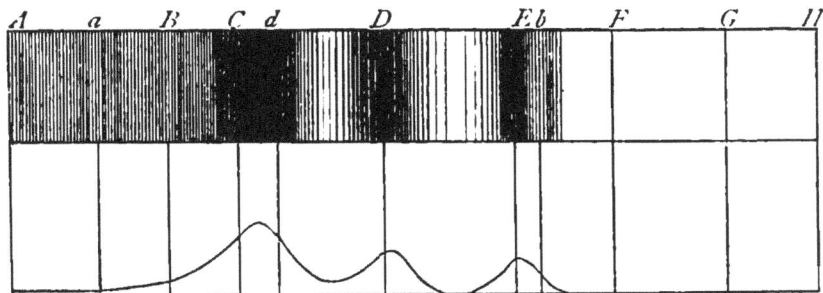

FIG. 5.

Absorption-spectra of organic colouring matters.

The graphic representation of an absorption spectrum may either be effected by shading the darkened parts

according to the degree of absorption, and leaving those
parts white which allow the rays to pass unchanged; or
by indicating the spectrum by a horizontal line and ex-
pressing the intensity of the absorption in each point by
means of co-ordinates. Fig. 5 shows, for the purpose of
illustration, the absorption-spectrum of blue cobalt-glass,

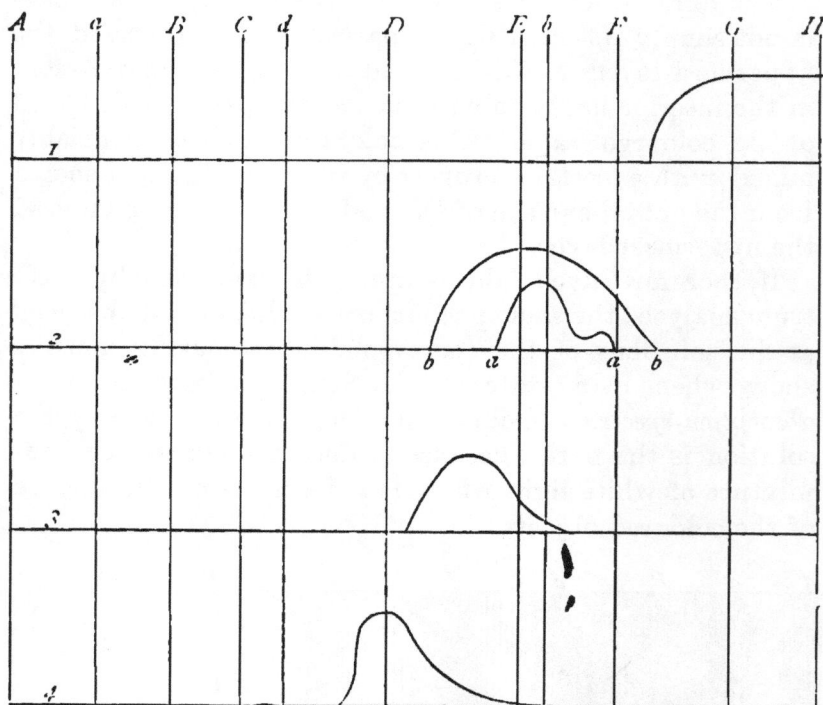

FIG. 6.

1. Picric acid.   2. Eosin : *a a*, dilute, *b b*, concentrated.
3. Magenta.   4. Methyl-violet.

represented according to each method. It will be seen
that cobalt-glass shows three bands of absorption, two of
which verge into each other.

The lines marked *A*, *B*, *C*, *a*, *d*, etc., indicate certain
prominent Frauenhofer's lines which are present in the

sun's spectrum, and which serve to determine relatively the positions of the different bands of absorption.

The following absorption-spectra of solutions of four well-known dye-stuffs—picric acid, eosin, magenta and methyl-violet (according to Vogel)—will serve as examples :—

Picric acid shows a total extinction of the blue side of the spectrum as far as $G \frac{2}{3} F$.*

Eosin in concentrated alcoholic solution absorbs chiefly the green, while in dilute solution it shows a dark band from $D \frac{3}{4} E$ to b. The absorption continues nearly as far as F, where it terminates in an indistinct narrow band.

Magenta is characterized by a very distinct band of absorption which, in dilute solutions, lies between D and E. In concentrated solutions it allows only the red rays to pass through.

By carefully considering the above examples, we arrive at the following deductions:—

The solutions of picric acid do not by any means possess the appearance of yellow because they allow only the yellow rays of the spectrum to pass through unabsorbed, but because they absorb those parts of the spectrum which are complementary to yellow; viz., the blue and the violet. The red, green and bluish-green rays are not absorbed any more than the yellow ones, but entering our eye they produce the impression of white. Thus the impression produced by picric acid is not a pure yellow, but consists of a yellow mixed with a considerable proportion of white.

In a similar manner, even concentrated solutions of the fine red eosin dye do not allow red light only to pass through; they only absorb that portion which lies between the yellowish-green and the pure blue, and allow part of the blue, the whole of the violet, red, orange and

* The expression $G \frac{2}{3} F$ serves to indicate that the spectrum is absorbed from the right end as far as $\frac{2}{3}$ of the space between G and F.

yellow, to pass through unchanged. The red colour of the
eosin solution is therefore caused by a suppression of the
green in the white, and not, as may have been supposed,
by a special property of allowing only red rays to pass
through.

The changes which the spectrum of a colouring matter
undergoes when dissolved in different neutral solvents are

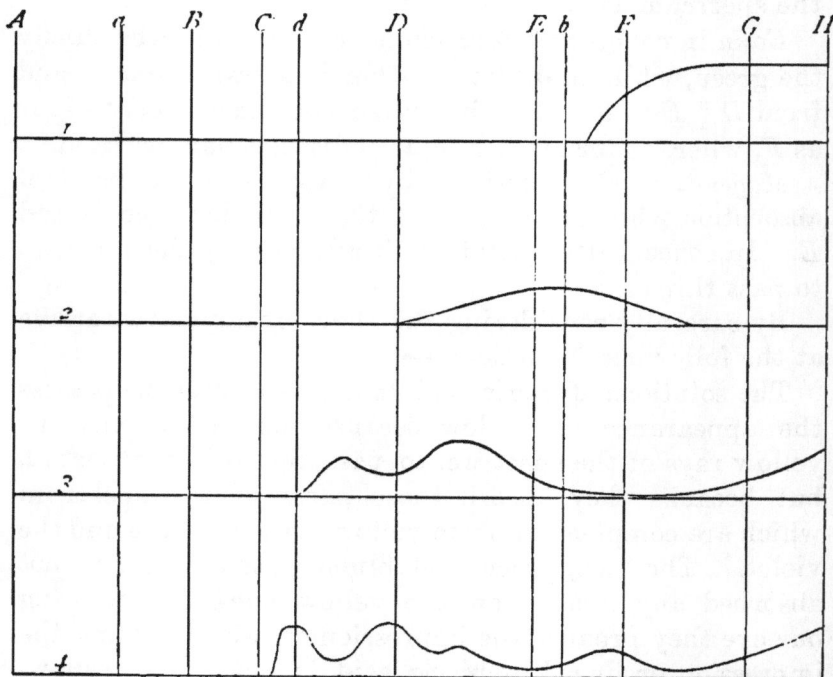

Fig. 7.

1. Alizarin in alcoholic solution. 2. The same with ammonia.
3. Alizarin in aqueous ammonia. 4. Alizarin in alcohol and
caustic potash.

generally very small. The colour lakes, however, and the
solutions of the colouring matters in acids or alkalies give
spectra which differ greatly from those of the colouring
matters themselves.

Fig. 7 gives an example (according to Vogel) of the

changes which the spectrum of alizarin undergoes when' converted into different salts. (2 and 4.) A comparison of the spectra 2 and 3 shows the effect of different solvents in the absorption.

Influence of concentration.—If the solution of a colouring matter is diluted by degrees, it will be seen that the lines and bands of absorption gradually become less distinct, until at length in very great dilution they become altogether invisible. If, on the other hand, the quantity of colouring matter is increased, either by adding more to the solution or by causing the light to pass through a deeper layer of the liquid, a point is ultimately arrived at when the liquid appears to be opaque and the spectrum is completely absorbed. This shows that the solutions of the colouring matters are neither perfectly transparent nor perfectly opaque for any kind of light rays. Dilute and concentrated solutions of one and the same colouring matter do not give the same spectrum, since the concentrated solutions always absorb a greater proportion of the rays, while the unabsorbed rays are considerably diminished in intensity. (See the spectrum of eosin, p. 14.) It is therefore necessary, in giving an exact description of the absorption-spectrum of a colouring matter, to state the concentration of the solution, and at the same time the depth of the layer through which the light has passed. A change in the spectrum is often accompanied by a change in the colour of the solution, which does not simply appear more saturated in a deeper layer, but has altogether changed colour. Chromic chloride offers a very striking example of this ; in dilute solution it appears green, while in concentrated solution it is red.

Solutions like that of chromic chloride, which possess an essentially different colour in dilute and concentrated solution, are called *dichroistic*. A solution of methyl-violet, when acidulated with hydrochloric acid, shows this property very distinctly.

C

In dyeing a series of shades with one and the same colouring matter, it is often seen that the light and the dark shades do not appear to have been produced with the same dye; the dark shades of a series of blues may, for instance, possess a distinct tint of red. This phenomenon is caused by a slight dichroism of the colouring matter used.

ABSORPTION-SPECTRA OF THE MOST PERFECT COLOURING MATTERS.

In order to obtain a knowledge of the value of colouring matters by means of their absorption-spectra, it is first of all necessary to know the nature of the absorption-spectra which the most perfect (purest) dyes give. Which rays, for instance, must a colouring matter absorb in order to impart to the material dyed with it the most brilliant yellow corresponding to the line D of the sun's spectrum? In order to decide this question, we will first study the impressions produced in our eye when two surfaces of equal size, the one white, the other coloured, are illuminated by lights of different colour.

If the white surface is illuminated by a sodium flame, *i.e.*, a light which contains only those rays of the spectrum which correspond to the line D, it will have the appearance of yellow, and will reflect only homogeneous yellow light. If the second surface had been dyed with a colouring matter, the absorption-spectrum of which contained but one bright line at D (Fig. 8), and were illuminated by white light (*e.g.*, electric light), it would absorb all the rays with the exception of those corresponding to the line D, and would, like the first surface, reflect only homogeneous yellow light.

If now the illumination by the sodium light and the electric light, respectively, were regulated so as to cause the same quantity of yellow rays to fall on each surface, they would, although in reality equal in intensity, not appear the same in our eye. This is due to the fact that the sensibility of our eye for objects which emit or reflect a limited number of rays decreases proportionately with the intensity of the light which is reflected into our eye from the surrounding objects. It is, for instance, well known that a grey disc of paper appears much lighter on a black ground than on a white one.

In the sodium light, all the objects surrounding the white surface appear of a yellow colour, which varies in intensity; but the intensity never exceeds that of the

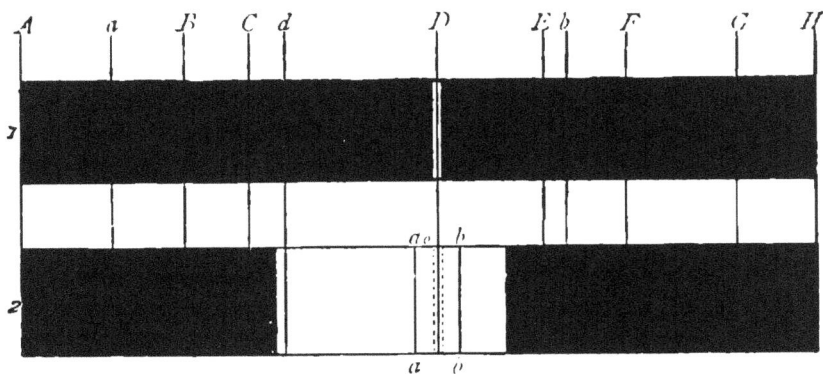

Fig. 8.

white object, because the latter reflects the whole of the yellow rays which strike it.

The yellow surface illuminated with the electric light absorbs, as we have seen before, all the rays with the exception of those corresponding to the line _D_ of the spectrum. The surrounding white objects reflect the whole of the rays, while the coloured ones also reflect a considerable proportion, and they will therefore all appear in a much brighter light than the dyed object, which only reflects a small quantity of light, and is therefore called a dull or dead yellow.

We know from the preceding that brightly coloured

objects reflect a number of different coloured rays which correspond to large portions of the spectrum, so that the reflected light contains a large proportion of the rays of light which strike it.

Adhering to our previous example, the most intense yellow will be produced in the following manner:—If the yellow space at D in the absorption-spectrum 1 (Fig. 8) is gradually enlarged on each side until it reaches the lines $a\ a$ and $b\ b$ in 2, besides pure yellow, a yellow containing a small proportion of red along with a yellow with a greenish tinge are allowed to pass through. The small quantities of red and green combined produce the impression of white in our eye, and we obtain therefore by this enlargement of the spectrum a large addition of yellow, but at the same time a small admixture of white. If now each side of the spectrum is gradually increased bit by bit, we shall find that the great increase of yellow obtained by the first enlargement gradually diminishes the more the ends of the spectrum deviate from the line D, while at the same time the green and red increase; i.e., the colour is mixed with more and more white.

If the spectrum is extended as far as the reddish-orange on the one side and a little further than the yellowish-green on the other, we only obtain a very small addition of yellow, whereas the white is considerably increased. A further enlargement of the spectrum would be of no use whatever, as the pure red and green would only combine to form white, and would thus diminish the fulness of the yellow. The most brilliant yellow corresponding to the line D would therefore be produced by a colouring matter which allows all the rays lying between reddish-orange and greenish-yellow to pass through unchanged, but absorbs all the other colours of the spectrum, as indicated in 2, Fig. 8.

In a like manner it can be shown that the most intense *green* must allow all the rays lying between yellowish-green and bluish-green to pass through, but must absorb the yellow, blue, orange, red and violet.

The finest *red* would be given by a colouring matter which allows red and orange, as far as pure yellow, and the rays lying between bluish-violet and violet, to pass through, and possesses, therefore, a spectrum consisting of two parts.

The most perfect *blue* would allow the blue end of the spectrum, beginning with bluish-green, to pass through.

This deduction only holds good for the primary colours, red, yellow, green, and blue, and must be somewhat modified if it is to be applied to intermediate shades. If, for instance, a colouring matter has to show an in-

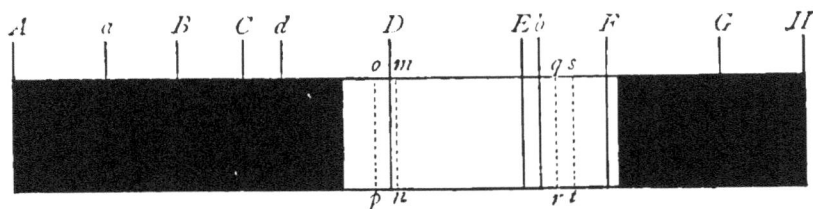

Fig 9.

tense yellowish-green, corresponding to that part of the spectrum which lies half-way between green and yellow, in the first place all the rays lying between pure yellow and green (from *m n* to *q r*) must pass through. The colour of the solution will then appear saturated in transmitted light, without any admixture of white. But the intensity of the colour can be considerably increased if the spectrum is enlarged on the one side as far as orange, on the other as far as blue. The colour adjoining the pure yellow is a yellow containing a very small proportion of red, while the colour adjoining the pure green is a green containing a very small proportion of blue. The small quantities of red and blue contained in the spaces *m n o p* and *q r s t* combine with the

equally small quantities of green and yellow contained in
the spaces $q \, r \, s \, t$ and $m \, n \, o \, p$ to form a small quantity of
white; but there still remains a considerable proportion
of green, which increases the intensity of the colour. If
the spectrum is gradually enlarged on both sides, the
yellowish-orange will first be neutralized by the greenish-
blue to white, while the red combines in a similar manner
with the green. The greenish-yellow is therefore not
intensified by a further increase of the spectrum. Hence
there is a limit to the perfection of a colouring matter.
It would be useless to try to obtain colouring matters
which impart shades to the dyed materials which correspond exactly to those of the spectrum, for if an object
has to appear of a brilliant colour, it must always reflect
some rays which combine to form in our eye the impression of white.

Mixtures of colouring matters.

When a new dye-stuff comes into the hands of a dyer,
he generally tries by experimental dyeing whether it is
available for the production of mixed shades; for the
most practised eye cannot foretell what shade will be
produced by mixing it with another colouring matter.
Thus two blues which cannot be distinguished from each
other in the pure state by the eye may yield when used
along with picric acid two shades which are as different
from each other as green from olive.

By an investigation of the green, for instance, which is
formed by indigo extract and picric acid, we obtain in
Fig. 10 an illustration of those parts of the spectrum
which are absorbed.

The solution of picric acid absorbs the violet and the
greater part of the blue rays, so that only those rays
which extend from the pure blue to the red end of the
spectrum can enter the solution of indigo extract, which

latter only allows the blue, bluish-green, and part of the
red to pass through unaltered, while the orange and the
yellow are almost completely absorbed, the green and
yellowish-green partially. By combining these two
spectra we obtain spectrum 3, which is characteristic for
the green produced by means of indigo extract and picric
acid. It consists of blue, bluish-green, green, and a
portion of the red.

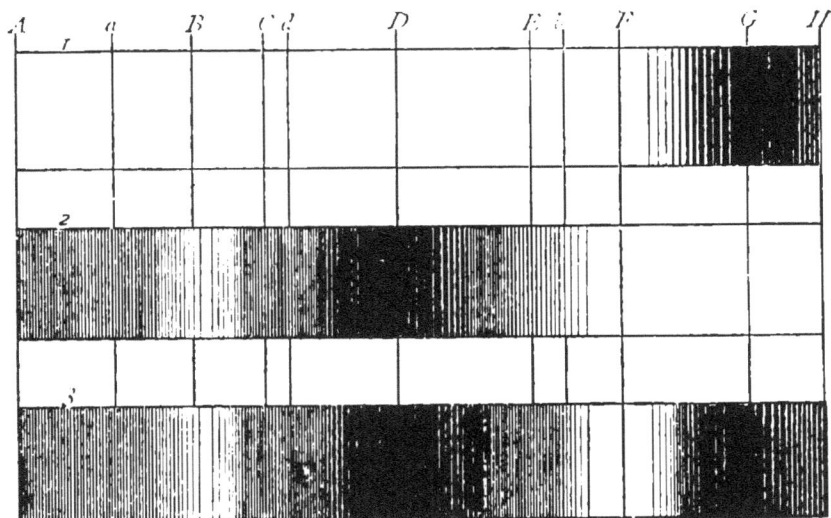

Fig. 10.

Indigo extract and picric acid do not, therefore, as
may have been supposed, yield a green because one
allows only blue rays, the other only yellow rays, to pass
through, but because they both allow green light to pass
through. This is shown a little more clearly in the
following example :—The spectrum 1 in Fig. 11 is that of
a pure blue colouring matter which would not yield a
green with a pure yellow colouring matter possessing the
absorption-spectrum 2, since all the rays would be absorbed, and the result would be a *black*.

If the spectra of these two colouring matters are enlarged on each side, the single colours will not undergo any appreciable change, only receiving an admixture of a certain proportion of white.

Nevertheless, a very fair green is obtained by mixing the two, which is shown in spectrum 3, Fig. 12.

A mixture of methyl-violet and Solid Green give a fine though not intense blue, the explanation of which is similar to that of the formation of green by mixing blue and yellow.

Violet yields with the yellow colouring matters dull

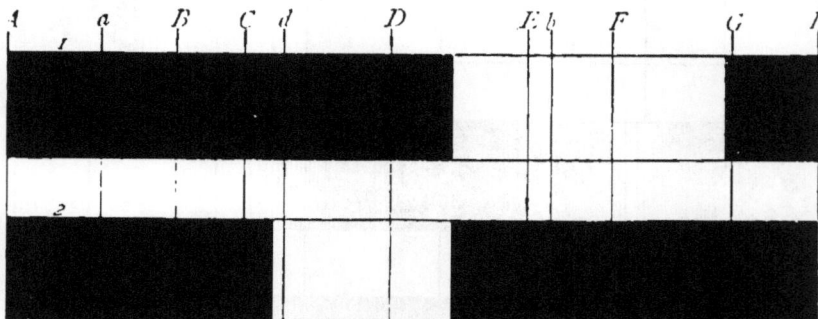

FIG. 11.

shades of green and yellowish-green, such as olives, moss-green, etc.

Of the more brilliant colouring matters there are probably no two which would fulfil the conditions expressed in Fig. 11, and would thus obstruct all the rays. This result can, however, easily be arrived at by using three colouring matters, which must be as different from each other as possible. Thus the combination of a red, a yellow and a blue colouring matter, when sufficiently concentrated, does not under ordinary circumstances allow any light to pass through, and can therefore be used for the production of *blacks*. This property is in fact made use of in dyeing, and not only black, but also an unlimited number of dark or dull shades of red, yellow and blue, comprising the greys and browns, can be produced in this manner. Light or dark greys are produced when small quantities of colouring matter are employed, and the proportions are chosen so that none of the three colours is present in excess. The colour inclines to brown, if red and yellow are in excess; while if the quantities of red and yellow are reduced, a bluish-grey is obtained.

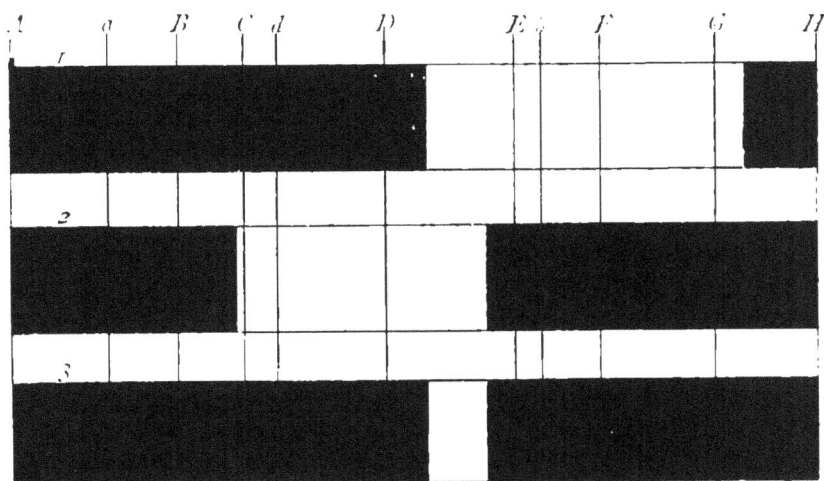

FIG. 12.

Fluorescence.

The light which enters a coloured liquid is not in every case simply divided into two parts, of which one is absorbed, while the other passes through. Some solutions do not only appear coloured by transmitted light but also in reflected light, in which case they appear to be rendered turbid by an extremely finely divided luminous solid in a state of suspension. Liquids which show this property are called *fluorescent*. A splendid example of fluorescence is shown, for instance, by an alcoholic, ammoniacal solution of Diazoresorufin. The liquid appears crimson by transmitted light, but in reflected light bright brick-red.

Fluorescein, some of the Eosins, Magdala-red, and Re-sorcin-blue also show a marked fluorescence when in solution.

The fluorescent properties of a colouring matter are shown to the best advantage on those fibres which possess most lustre. Silk shows it best; wool will also show it under certain circumstances; but cotton cannot be made fluorescent, because the rough surface of the fibre disperses, and thus thoroughly mixes the transmitted light and the light of fluorescence.

The Spectroscope as a means of detecting colouring matters.

The absorption-spectra of the colouring matters can in some cases be made use of for their detection, especially when they contain characteristic lines of absorption, either in neutral, acid or alkaline solution. Thus, alizarin can be recognised by the bands shown in the spectroscope, by allowing the light to pass through its alkaline alcoholic solution. Reliable methods have indeed been devised for detecting and distinguishing from each other certain colouring matters in this way, but the field has not yet been systematically worked. Colouring matters which are closely allied to each other in chemical constitution, and are therefore difficult to distinguish from each other by their chemical reactions, generally give very similar spectra; on the other hand, colouring matters which possess very different spectra may also be easily distinguished from each other by chemical means. Analysis by means of the absorption-spectrum, which has become of great importance for many branches of science and practice, is therefore only of minor importance to the colour chemist, and it is only in a few cases that it renders him important services.

The use of the spectroscope has also been suggested by Schoop for the quantitative determination of the coal-tar colours. (See *Journ. Soc. Dyers and Col.*, 1886, p. 71.)

GENERAL CHEMICAL PROPERTIES OF THE COLOURING MATTERS.

The term "colouring matter," in the strict sense of the word, embraces all those coloured substances which in a state of fine division are capable of imparting their own colour to other objects. The term is, however, frequently applied to other bodies, which of themselves possess little or no colour, but can be transformed into colouring matters proper by the action of bases or acids. The chemical denomination has indeed become so usual that it has in many cases almost completely replaced the former. Thus alizarin is called a colouring matter, although it possesses in the free state only a dull yellow colour, whereas the beautiful red combination of alizarin with alumina is generally known as a *Colour-lake*, and not as a colouring matter proper.

Chemical constitution.—This chapter will chiefly be devoted to the consideration of those chemical properties of the colouring matters which are of importance in their application, and especially in their fixation to the fibres ; and since the terms " chemical constitution " and " constitutional formula " will frequently be made use of in this and the following chapters, it may be well to give a thorough explanation of their meaning.

According to the chemical theory at present generally adopted, the smallest particles of the elements which can enter into chemical combination are called *atoms*. These atoms combine with each other according to definite laws, in greater or smaller numbers, to form *molecules*, which are simultaneously the smallest particles of matter capable of existing in the free state. Thus, each molecule of magenta is composed of atoms of carbon, hydrogen, nitrogen and chlorine.

The abridged chemical notation by which the number

of these atoms and the proportion in which they are contained in the molecule are stated, is called an *empirical formula*. Thus, the fact that the molecule of magenta consists of twenty atoms of carbon (C), twenty atoms of hydrogen (H), three atoms of nitrogen (N), and one atom of chlorine (Cl), is expressed by the empirical formula, $C_{20}H_{20}N_3Cl$.

By a thorough investigation of the chemical properties of these bodies, their modes of formation, products of decomposition, and the compounds they are capable of forming, modern chemistry has procured a deeper insight into their interior structure. But although we have no definite idea as to the actual relative position of the atoms in the molecule, we are nevertheless enabled to lay down a scheme in which the atoms given by empirical formulæ are arranged in such a manner as to show the whole chemical nature of the compound. A scheme of this kind, which simply serves to express our knowledge of the compound to which it relates, is called a *constitutional formula*. The determination of the constitutional formulæ of bodies like the organic colouring matters, the composition of which is generally very complicated, is often a difficult matter, and it is therefore not to be wondered at that there are still many colouring matters the constitution of which is unknown.

In the constitutional formulæ certain groups of elements invariably indicate similar chemical properties. Thus we know that bodies which contain one atom of nitrogen and two of hydrogen combined, as NH_2, all have the property of combining with acids. On the other hand, each individual chemical property is not a function of the arrangement of the total number of atoms, but only of certain groups of atoms.

In order to obtain a general idea of the chemical nature of the colouring matters, it will be found convenient to divide these groups of atoms into two classes; viz. (1), those which cause the colour of the compound; and (2) those on which the acid or basic nature of the compound depends.

1. The groups which cause the colour of a compound are knows as *chromophorous* or colour-bearing groups. They can often be discovered by carrying out certain reactions with the compound, and observing in which cases it loses its characteristic colour. It is then necessary to find out which groups have undergone a chemical change. If this has taken place with one group only, then that group is the chromophorous one; while if several groups have been affected, it is in most cases possible to detect the chromophorous group by a comparison of the results of several reactions.

Thus we know that the yellow colour of picric acid, $C_6H_2(NO_2)_3OH$, is due to the presence of the three nitro (NO_2) groups; because if this colouring matter is reduced by means of tin and hydrochloric acid, it is decolourised, and converted into triamidophenol, $C_6H_2(NH_2)_3OH$, in which the nitro groups of the original compound have been replaced by amido groups.

In a similar manner it has been ascertained that the chromophorous group contained in all the azo-colours, consists of two atoms of nitrogen, combined in the following manner:— $N:N$—. But although it may have been proved that a group of atoms has caused the colour of a certain compound, it must not be concluded that this group is chromophorous in every case, and that therefore every compound that contains it must possess a similar colour. Generally this is the case with bodies of similar constitution, but although, for instance, most nitro-compounds are coloured yellow or orange, there are others which are colourless. The properties of the chromophorous group are of special interest to the dyer when he wishes to bring about a temporary or permanent decolourisation of the colouring matter, such as is necessary in the preparation of "vats" or in the production of "discharges" (in calico-printing).

Salt-forming groups.—The groups of atoms which cause the colouring matters to form salts can also easily be recognised in the constitutional formulæ, and by a knowledge of them we can see at a glance their relation to the animal and vegetable fibres. For this purpose they may be suitably classified, according to their property of combining with acids or bases, or with neither, as *basic colouring matters, acid colouring matters*, and indifferent or *neutral colouring matters*.

The *basic colouring matters* are always used in dyeing in the form of their salts—*i.e.*, of their compounds with mineral or organic acids. Colouring matters possessing a distinct basic character, which lose their colour in coming in contact with acids, cannot be used in dyeing, as will be shown further on. The only exceptions to this rule are some very weak bases, which are absorbed by the fibre as such and not in the form of their salts. Amidoazobenzene, $C_6H_5 - N : N - C_6H_4 \cdot NH_2$, a colouring matter which was formerly sold and used under the name of "Aniline-yellow," furnishes an example of this kind. Its salts are decomposed by water; if, therefore, silk is dyed with an acidulated solution of its hydrochloride, and is then washed with water, the free base only remains on the fibre. If, however, the silk undergoes a subsequent treatment with dilute acids (*e.g.*, in brightening), the colouring matter evinces a neutral reaction and remains unchanged. The free colour-bases are for the most part colourless or only slightly coloured.

All the known basic colouring matters contain nitrogen, and it is to the presence of these nitrogen atoms that they owe their basic properties. They are all derived from the type ammonia :—

$$N \begin{cases} H \\ H \\ H \end{cases}$$

By replacing one, two or all three of the hydrogen atoms in this compound by organic radicals, the *amines* or amido-compounds are formed. These are of three kinds. The *primary* amines contain two replaceable hydrogen atoms, the *secondary* amines one, while in the *tertiary* amines all the hydrogen atoms are replaced by organic radicals. The following may serve as examples :—

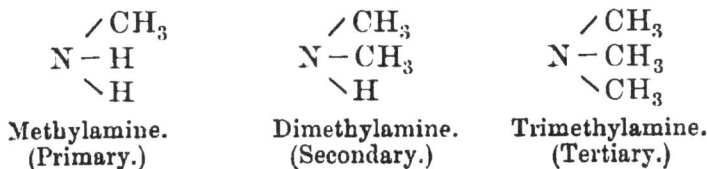

$$
N \begin{array}{l} \diagup CH_3 \\ - H \\ \diagdown H \end{array}
\qquad
N \begin{array}{l} \diagup CH_3 \\ - CH_3 \\ \diagdown H \end{array}
\qquad
N \begin{array}{l} \diagup CH_3 \\ - CH_3 \\ \diagdown CH_3 \end{array}
$$

Methylamine.	Dimethylamine.	Trimethylamine.
(Primary.)	(Secondary.)	(Tertiary.)

In place of methyl (CH_3) any other organic radical can be substituted ; *e.g.*, ethyl (C_2H_5), phenyl (C_6H_5), benzyl (C_7H_7), etc.

In the amines, therefore, the nitrogen is always in direct combination with carbon and hydrogen or with three carbon atoms. The group NH which is contained in the secondary amines is known as the Imido group.

The colour-bases may either contain one or several amido groups. Of two or more colour-bases possessing similar constitution, the one containing the largest number of amido groups will combine most easily with acids. The number of equivalents of an acid with which the base will combine is, however, limited according to the number of hydrogen atoms it contains in direct combination with the nitrogen atoms. Thus, rosaniline, $C_{20}H_{19}N_3$, contains six such hydrogen atoms, and it yields, besides the salts containing one equivalent of acid, such as ordinary magenta, $C_{20}H_{19}N_3 \cdot HCl$, another series of salts, of which $C_{20}H_{19}N_3 \cdot 3HCl$ would be a representative.

The basic properties of the primary amines, and also of the colouring matters containing primary amines, are much more marked than those of the secondary or tertiary amines. They are also considerably diminished by the introduction of electro-negative elements or groups, such as Chlorine, Bromine, Iodine or Nitro groups. Thus aniline, $NH_2C_6H_5$, which is a primary amine, is a strong base; diphenylamine, $NH(C_6H_5)_2$, which is a secondary amine, forms salts which are easily decomposed; while hexanitrodiphenylamine, $N(C_6H_2[NO_2]_3)_2H$, is no base at all, but, on the contrary, a strong acid. Its salts are known in commerce as Aurantia.

The salts of those colour-bases which contain unsubstituted amido groups are generally soluble in water, but if the hydrogen of these groups is partially or wholly replaced by organic radicals, the solubility decreases proportionately. Thus, magenta, which may be represented as pararosaniline hydrochloride, which has the formula—

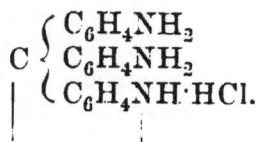

$$C \begin{cases} C_6H_4NH_2 \\ C_6H_4NH_2 \\ C_6H_4NH \cdot HCl. \end{cases}$$

and contains two amido and one imido group, is comparatively easily soluble in water, whilst aniline-blue or the hydrochloride of triphenylrosaniline—

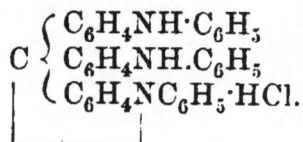

$$C \begin{cases} C_6H_4NH \cdot C_6H_5 \\ C_6H_4NH.C_6H_5 \\ C_6H_4NC_6H_5 \cdot HCl. \end{cases}$$

is insoluble in water.

Most of the colour-bases can be transformed into soluble acid colouring matters by the action of sulphuric acid. (*vide* Sulphonic Acids).

The *acid colouring matters* contain, like all acids, hydrogen atoms which can easily be replaced by metals; in other words, they possess the property of combining with bases to form salts with the simultaneous evolution of water. Thus, picric acid, $C_6H_2(NO_2)_3OH$, contains a hydrogen atom which can be replaced by metals:—

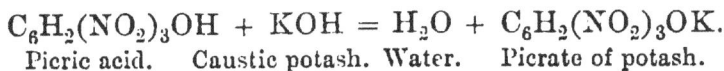

$$C_6H_2(NO_2)_3OH + KOH = H_2O + C_6H_2(NO_2)_3OK.$$
Picric acid.   Caustic potash.   Water.   Picrate of potash.

As will be seen from the above example, not all the hydrogen atoms contained in an organic compound are replaceable by metals, but only those which are contained in certain groups, and it is to these that the compound owes its acid character. They present themselves in the form of the *hydroxyl* (OH) group or of the *sulphonic acid* group (SO_3H); sometimes, also, as in aurantia, in the form of the imido group (NH), or as in the yellow obtained from salicylic acid and benzidine, in the form of the carboxyl group (COOH). All those colouring matters which owe their acid properties to the presence of hydroxyl groups are only weak acids. Thus the salts of fluorescein, $C_{20}H_{10}O_3(OH)_2$, are decomposed by acetic acid. Their acid properties are, however, considerably intensified by the introduction of chlorine, bromine, iodine, or nitro groups. It is well known that picric acid, $C_6H_2(NO_2)_3OH$, is a much stronger acid than phenol, C_6H_5OH. Again, tetra-brom-fluorescein, or eosin, $C_{20}H_8Br_4O_3(OH)_2$, is a much stronger acid than fluorescein. The colouring matters of this group are either insoluble or only sparingly soluble in water. Alizarin dissolves in 3,000 pts. of boiling water, picric acid in 86 pts. of water at 15° C. The substitution products containing Chlorine, Bromine, Iodine, etc., are still less soluble. Thus, whereas fluorescein can be dissolved in a large quantity of boiling water, eosin (free) is completely insoluble.

On the other hand, most of these colouring matters are easily soluble in dilute alkaline solutions, or in other words, their alkali salts are soluble in water. The solubility of these salts in water lies in a direct ratio to the number of hydroxyl groups; but if the hydrogen of these

D

groups is replaced by organic radicals, such as methyl, ethyl, benzyl, etc., the compounds are again rendered less soluble or even insoluble in water. Thus, the ethyl ether of eosin, $C_{20}H_6Br_4O_3(OH)(OC_2H_5)$, yields a potassium salt which is insoluble in water.

Many of the colouring matters of this class yield insoluble compounds with the metallic hydrates, such as those of chromium, iron, aluminium, etc., which are known as *colour lakes*, or simply as *lakes*.

SULPHONIC ACIDS.—The sulphonic acids can either be derivatives of the basic or of the phenol colouring matters. They all contain one or more sulpho groups, and are characterised by their strongly acid properties. They can either be obtained by the action of sulphuric acid on the colouring matters themselves; *e.g.* :—

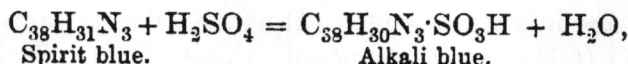

$$C_{38}H_{31}N_3 + H_2SO_4 = C_{38}H_{30}N_3 \cdot SO_3H + H_2O,$$

Spirit blue.   Alkali blue.

or by the direct transformation of the sulphonic acids into colouring matters :—

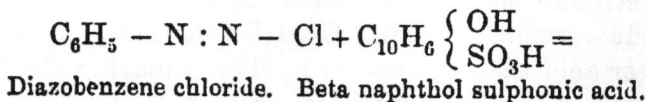

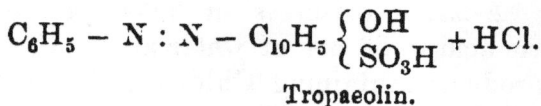

$$C_6H_5 - N : N - Cl + C_{10}H_6 \begin{cases} OH \\ SO_3H \end{cases} =$$

Diazobenzene chloride.   Beta naphthol sulphonic acid.

$$C_6H_5 - N : N - C_{10}H_5 \begin{cases} OH \\ SO_3H \end{cases} + HCl.$$

Tropaeolin.

The sulpho group is not chromophorous; it has so little influence on the colour of the compound that the original and the sulpho derivative are generally of exactly the same colour. The sulphonic acids of the colouring matters are of great importance in practice, as they are generally soluble in water and can be dyed in acid baths. Colouring matters which are volatile in steaming can be rendered non-volatile by transforming them into their sulphonic acids. The salts of the dyes containing sulpho groups are much more soluble in water than those of the dyes containing only hydroxyl groups. They either possess a dull, indefinite colour (Alkali-blue, etc.), or they show their characteristic colour (Azo-dyes).

Some colouring matters belonging to this class also yield lakes with metallic hydrates.

Neutral colouring matters.—Very few artificial colouring matters possess neither acid nor basic properties. As an example, we may mention the artificial indigo, obtained from propiolic acid.

METHODS OF DISSOLVING THE COLOURING MATTERS.

The colouring matters are generally added to the dye-bath in a state of solution, sometimes also in the form of a paste (*e.g.*, alizarin paste), which latter does not contain the colouring matter in solution, but in a very fine state of division. As solvents, only those liquids which are miscible with water can be used for dyeing purposes; petroleum, benzene, etc., are therefore excluded. The most usual solvents are water, alcohol (methylated spirits), and acetic acid.

The water which is used for dissolving should be as pure as possible, especially with respect to lime salts, since the latter tend to precipitate part of the colouring matter. This evil can, however, be remedied by neutralizing the water with a few drops of acetic acid.

The solutions are usually prepared hot, and are then either filtered or decanted in order to retain insoluble impurities, or particles of the dye which may not have been dissolved. If this operation is omitted, the goods are liable to become spotted or stained.

Many colouring matters dissolve easily in hot water,

but are so sparingly soluble in cold water that they separate out again on cooling. It is advisable, in making large quantities of such solutions for use, to add methylated spirits or acetic acid.

The term " soluble in spirit " implies at the same time that the colouring matter in question is insoluble in water. There are, however, many colouring matters which dissolve easily in both these solvents. The use of the colouring matters soluble in spirit is very limited, owing to the fact that their application is not only more expensive, but also more difficult than that of the colouring matters soluble in water.

The aqueous solutions of the coal-tar colours are often used as *coloured inks*, for the preparation of which 1 pt. of colouring matter is dissolved in about 50 pts. of water, with the addition of a little gum. "Aniline inks," which are made in pretty concentrated solution, show after drying on paper the characteristic metallic lustre of the colouring matter itself. Thus, marks produced with strong violet ink show a green reflex. Copying inks are produced by adding, besides gum, glycerin to the solution of the colouring matter. Important manuscripts or documents should not be written with these inks, as the colours gradually fade, especially when exposed to light.

The methylated spirits used for dissolving should be as concentrated as possible, and at the same time free from impurities, especially such as aldehyde and acetone, which act very injuriously on certain dyes. It is well known that magenta, for instance, when treated with aldehyde is transformed into a fine blue colouring matter. The small quantity of aldehyde which might be contained in the methylated spirits would, however, suffice to convert the fine red shade of magenta into a violet. Other colouring matters, such as violet soluble in spirit, are similarly affected. This reaction can, on the other hand, be employed as a qualitative test for impurities in methylated spirits. In order to carry out the test, small quantities of magenta are dissolved in pure methylated spirits and in the methylated spirits to be tested. The two liquids are then brought to the same intensity of colour by diluting one of them, and the shades are compared. If the spirits in question are of exactly the same colour as the pure spirits, this is a sign that it contains no acetone or aldehyde; but if it is coloured violet, it is impure.

Open vessels should not be used for dissolving colouring matter in methylated spirits, on account of the loss which would be incurred by evaporation. The most useful form of vessel for this purpose is a small copper boiler, having a capacity of several litres, and provided with a lid which can be screwed down like that of a Papin's digester. The lid should be provided with a safety-valve. The boiler containing the colouring matter and the methylated spirits is placed in boiling water, when, taking into account that pure alcohol boils at 78·4° C., water at 100°, a considerable pressure will be formed in the interior. (For absolute alcohol the presure would represent somewhat more than two atmospheres.) Solution is effected much more rapidly in this way than under the ordinary atmospheric pressure. After some time, the boiler is taken out, allowed to cool, and the clear liquid is decanted from any insoluble residue. Alcoholic solutions, prepared in this or other ways, should not allow the colouring matter to precipitate on being diluted with a small quantity of water, otherwise it would not be evenly divided throughout the dye-bath. Some colouring matters, e.g., blue soluble in spirit, are of necessity precipitated, but in such a fine state of division that they are evenly assimilated by the fibre.

Colouring matters which require large quantities of methylated spirits for their solution are objectionable in many respects, and are therefore not much used in practice. For besides materially increasing the cost of dyeing, a large quantity of spirits is apt to coagulate * the " boiled-off liquor," and thus render it useless in dyeing.

If a colouring matter is only sparingly soluble in water and in strong methylated spirits, it is advisable to try whether it will not dissolve in a mixture of the two liquids. There are, in fact, colouring matters which dissolve much more easily in a 50 per cent. solution of methylated spirits than in either water or spirits alone.

The alcoholic solutions of the coal-tar colours are used to some extent in the preparation of brightly coloured *varnishes*. These are prepared by dissolving different resins (shellac, copal, dammar, sandarac, etc.) in strong spirits of wine, and adding to the concentrated solution sufficient colouring matter to produce the required shade.

Acetic acid, $CH_3 \cdot COOH$, acts in many cases as an effective solvent for the artificial colouring matters.

Glycerin, $C_3H_5(OH)_3$, dissolves many of the coal-tar colours in large quantities, and is often used in the mixing of printer's colours. What makes it so valuable an agent in this respect is its property of simultaneously dissolving many other constituents of the printer's colours, especially albumen, gum, arsenic, etc. Glycerin is also sometimes used, on account of its viscosity, in the preparation of colour-pastes.

Solutions of *gum* and *glue* can also be used as solvents for many aniline dyes. The solutions may be prepared in the following manner:—

The solution of the colouring matter in alcohol or methylated spirits is added to a concentrated solution of gum-arabic, and the whole is well stirred. The colouring matter passes from the alcoholic solution into the gum, from which latter it is not precipitated by a subsequent dilution with water.

Another method is to dissolve glue in acetic acid at 10 to

* The sericine is precipitated. See p. 41.

12° Tw. and add to this solution the finely-pulverized dye. The mass is then well stirred and heated in a closed vessel, until a sample taken out and treated with water dissolves without any residue. The solution obtained in this manner can be allowed to solidify, in which state it can be kept for future use.

Very few of the coal-tar colours are soluble in *petroleum*, *benzene*, and *ether*, but many organic liquids of higher boiling-points, such as *nitrobenzene, aniline oil, fusel oil, and carbolic acid*, act as eminent solvents.

In the *neutral oils* and *fats*, most of the aniline dyes are insoluble (methyl-violet is an exception). The basic colouring matters will, however, dissolve readily in *fatty acids*, or in oils containing fatty acids.

Fatty compounds of the aniline dyes can be prepared by adding the free bases to oleïc or stearic acid. These compounds are used for colouring oils, fats, oil-varnishes, etc. Another method of preparing fatty compounds of the basic colouring matters is to precipitate the solution of the dye with a solution of soap. Thus, magenta and soap would yield by double decomposition sodium chloride and a fatty compound of rosaniline.

Turkey-red oil is the ammonia or soda-soap of castor oil, and acts, therefore, as a solvent for aniline dyes.

For the preparation of coloured cosmetics or candles, special care should be taken in selecting colouring matters which are absolutely free from poisonous constituents.

DYEING WITH COAL-TAR COLOURS.

Chemistry of the fibres.

Although the question as to whether the colouring matters enter into a chemical or a mechanical combination with the fibres still remains undecided, it is nevertheless certain that the relation of the colouring matters to the fibres does not only depend upon the physical structure of the latter, but also to a great extent on their chemical composition It is therefore necessary to say something here on the chemistry of the fibres.

Of the numerous fibres used in the textile industries, three—viz., silk, wool, and cotton—form the chief bulk, and it will suffice to give a description of these as representatives of the textile fibres in general.

SILK.—Silk does not consist of one homogeneous substance. It is possible, by means of the microscope alone, to discern two, sometimes three different layers, which differ from each other chiefly in their chemical properties. In the natural state the silk fibre proper is enveloped in a coating (silk-gum), which in the case of yellow silk contains the colouring matter.

The chemical compound of which silk consists is known as fibroine, and somewhat resembles in its properties keratine, the substance of horn, wool, hair, etc. The relative quantities of silk-gum and actual fibre contained in the raw silk vary considerably, according to origin, quality, etc.; but in practice 18–27 per cent. of silk-gum and 73–82 per cent. of fibroine are generally calculated upon.

Neutral solvents have no action on fibroine. Prolonged heating with water, even under pressure, will not render it soluble. By caustic alkalies it is easily dissolved, and is attached by the same even in dilute solutions. Ammonia, however, as well as the carbonates and sulphites of the alkalies, and neutral soaps, have very little action. Concentrated mineral acids, such as sulphuric acid, nitric acid and hydrochloric acid, dissolve the silk fibre, while it is deeply attacked by chlorine, bromine, and iodine. Sulphurous acid has little action, and can therefore be used for bleaching silk.

The external covering of raw silk, which is also known as " silk-gum " or " silk-glue," can be removed by boiling the silk for several hours in water under pressure. The solution obtained in this manner solidifies on cooling to a gelatinous mass, which when evaporated to dryness yields a product closely allied to glue. The solutions of silk-glue are precipitated by alcohol; tannin substances also produce insoluble precipitates in them.

Both fibroine and sericine (silk-glue) consist of carbon, nitrogen, hydrogen, and oxygen. They do not contain any sulphur. Raw silk leaves on ignition from 1 to 1·5 per cent. of ash, consisting chiefly of carbonate of lime. The percentage of ash is often considerably increased by the use of hard water in softening the cocoons, and is in many cases the cause of unfavourable results in dyeing, since insoluble lime lakes are formed, which render the shade of the material "cloudy." The injurious effect of lime in the fibre is most prominent when the oxalates of the colouring matters are used in dyeing, owing to the fact that insoluble oxalate of lime is formed in the fibre, and thus prevents the dye being absorbed evenly. Most of the lime can, however, be got rid of by treating the silk, previous to dyeing, with very dilute lukewarm hydrochloric acid.

Silk is very seldom dyed as raw silk; in the majority of cases it is first "discharged," or deprived to a greater or less extent of its external covering, the silk-glue. By this means it becomes more glossy, and a better feel is imparted to it, while at the same time it loses in weight.

The *stripping* or discharging of silk is effected by boiling the raw silk in strong soap solutions. On the large scale this is generally carried out in two operations. In the first soap-bath, the temperature of which is not quite 100° C., the hanks of silk are worked until the silk-glue swells up and falls from the fibre.

In the next operation it is sewn up in linen bags and boiled in a fresh soap-bath, until it has acquired its proper feel and handle.

This second bath can be used again for ungumming a fresh portion of silk, and is used lastl in the form of *boiled-off liquor* as an addition to the dye-bath (in silk dyeing).

The loss in weight in these two operations varies from 18 to 27 per cent. It is sometimes desirable to transform raw silk into a glossy and " scroop " product without involving such a considerable loss in weight. This is done by treating the silk with water containing tartar and some sulphuric acid, heated nearly to the boil. By this method the silk is only partially discharged, and the loss in weight is only from 4 to 8 per cent. The product is known as *souple*, and lies, with respect to its properties, between raw and discharged silk.

It follows from this that raw silk, souple and discharged silk, must be acted upon differently by chemical agents, and must therefore not all be treated in a like manner. Thus, boiled-off silk, which consists of almost pure fibroine, can be dyed in very hot acidulated or slightly alkaline baths ; while raw silk and souple will not stand this treatment, on account of the silk-glue they contain.

WOOL.—Wool is never dyed in the raw state—*i.e.*, as it comes from the sheep's back—but is always previously washed and scoured. By these operations the " suint " or " yolk " (a substance consisting of cholesterin, the alkali salts of certain organic acids, earthy impurities, etc.) is removed, thus leaving behind almost pure wool. Wool consists of a substance chemically known as keratine, an albuminoid body, which is also the chief constituent of hair, horn, feathers, etc. It contains, besides carbon, nitrogen, hydrogen, and oxygen, a certain percentage of sulphur, which causes it to differ considerably from silk in its properties.

Wool is completely dissolved by hot solutions of caustic alkalies. It is not much affected by dilute acids, but concentrated mineral acids attack the fibre ; in strong sulphuric acid it dissolves readily at 100°, while strong nitric acid dissolves it with a yellow colour, producing xanthoproteïc acid. When boiled with dilute acids, wool absorbs a considerable percentage of acid, which can only be removed by neutralising with an alkali or by long-continued boiling with water. Thus wool which had been boiled in water containing 10 per cent. of sulphuric acid (of the weight of the wool) had to be boiled nine times in succession for an hour each time, in fresh distilled water, before all the acid could be removed. A small proportion of the acid appears to be neutralized by some basic constituent or constituents of the fibre. It is interesting to note that wool treated in this manner (one might even say mordanted) can be dyed a full shade in *neutral* solutions of the acid coal-tar colours. It should be borne in mind that the sulphurous acid which is used almost exclusively in the bleaching of wool also adheres to the fibre very tenaciously after the operation is over. After stoving, the material should first be rinsed in a weak solution of soda and then thoroughly washed with water containing hydrogen peroxide. If this latter operation is omitted, and the material is subsequently dyed with an aniline dye, the sulphurous acid retained by the fibre may gradually reduce and decolourise the colouring matter. This is sometimes seen very distinctly in certain fabrics in the places where dyed and bleached yarns cross each other.

COTTON.—Clean and well-bleached cotton consists, like most other perfectly bleached vegetable fibres, of almost pure cellulose. Cellulose consists of carbon, hydrogen, and oxygen in the proportions expressed by the formula, $(C_6H_{10}O_5)n$. The only portion of the cotton fibre which does not consist of cellulose is the extremely thin membrane or cuticle which envelopes it, and to which it owes its tenacity in a great measure. The chemical composition of this cuticle is not known. If it is destroyed, the fibre falls to a powder. Cotton is dissolved by concentrated sulphuric acid; nitric acid converts it into gun-cotton ; while hydrochloric acid disintegrates the fibre. Caustic alkalies and dilute solutions of bleaching powder have very little action on the cotton fibre.

For the dyer, the most important chemical properties of the fibres are those which bear directly on the colouring matters themselves, or on the additions made to the dye-bath. In dyeing, it is not only necessary to see that the colours produced on the different materials are as full and as fast as possible, but also that the fibre should not lose in strength, and in the case of wool or silk, that it should not part with any of its lustre or pliability. After dyeing, wool should still feel soft, and not harsh or "hask"; while silk should retain its peculiar feel, and when compressed or twisted strongly, should emit a peculiar crackling sound.

From these short notes on the chemistry of the fibres, it will be seen that strongly acid or alkaline baths cannot be used indiscriminately for any kind of fibre without some-times incurring considerable injury, and it is therefore advisable to observe the following rules in dyeing.

Action of acids and alkalies on the fibres.—Wool and silk will stand the action of dilute acids without injury, and as long as strong solutions of the mineral acids are not made use of, it is not to be feared that the fibres will undergo any change when dyed in acid baths. Indeed, silk only acquires its peculiar crackling or "scroop" feel after a treatment with an acid ; after being dyed and washed, it is therefore almost invariably passed through a weak acid bath, an operation known as developing (brightening). Acetic, tartaric, or citric acids are generally used for this purpose ; but it is preferable to use one of the latter two, as the acetic acid absorbed by the fibre gradually eva-porates, and the silk then loses its peculiar feel.

Hot solutions of mineral acids, even when very dilute, have an injurious effect on *cotton* and other vegetable fibres ; they weaken or "tender" the material. If it is necessary to dye cotton in an acid bath, the material should be thoroughly washed subsequently, for if a mineral acid, even if very dilute, is allowed to dry on the fibre, the latter will be gradually destroyed.

The *animal fibres* are very easily attacked by caustic and even by carbonated alkalies. An 8 per cent. solution of caustic soda will dissolve them completely. Even dilute soda solutions, which are sometimes used for discharging certain kinds of silk, have a destructive action on the fibre if allowed to lie in contact with it for a length of time. The use of baths which contain either caustic or carbonated alkalies should therefore be avoided as much as possible in the dyeing of wool or silk.

On the other hand, the alkalies have not an injurious effect on cotton. Even concentrated solutions of the caustic alkalies do not tender the fibre, but only cause it to contract and become more dense (mercerised cotton).

The Relations of the Fibres to Colouring Matters.

Taking exception to artificial indigo, aniline-black, and one or two other colouring matters, the coal-tar colours can be divided into two classes with respect to their relation to the fibres :—

1. The colouring matter is absorbed directly from its solution by the fibre. In this case the fibre is said to be *substantively* dyed, and the colouring matter is called a *substantive* colouring matter.

2. The fibre does not combine directly with the colouring matter, and must first be charged with metallic salts or hydrates (*mordanted*), or prepared in some other way before it will receive the colour. These are known as *adjective* colouring matters.

Substantive dyeing.—The animal fibres possess great affinity for most of the coal-tar colours, and in many cases they absorb them so completely that the liquid is rendered colourless. Many colouring matters are taken up by wool and silk from an acid bath much more readily than from a neutral bath. The bath is rendered acid (soured) by the addition of acetic acid, tartaric acid or sulphuric acid. When sulphuric acid is used, Glauber's salts (sodium sulphate) are usually added simultaneously, and the two combine to form sodium bisulphate, which has less action on the fibre than free sulphuric acid. Some dyers prefer to add the sodium bisulphate as such, instead of forming it in the dyebath. In this case, care should be taken to see that the commercial product is free from nitric acid, which has a very injurious action on many colouring matters. In wool-dyeing, sodium sulphate or magnesium sulphate are frequently added to the dye-bath, in order to reduce the solubility of the colouring matter, and thus to obtain more even and faster colours.

Silk absorbs many colouring matters so rapidly that the colour produced is uneven or "cloudy." In order to prevent this, silk is usually dyed in a weak soap-bath. Neutral or alkaline soap-baths are made up with Marseilles or good olive-oil soap; while for acid soap-baths, boiled-off liquor is used, along with acetic or sulphuric acid.

Cotton evinces very little affinity for most of the coal-tar colours. Thus picric acid, naphthol yellow, the azo-scarlets, and most other so-called acid dyes, cannot be fixed permanently in cotton; the basic dyes can be fixed by the aid of mordants (tannin), while chrysamine, Congo red, and other dyes of this class, will dye cotton directly in an alkaline bath. Gun-cotton, collodion, and vegetable parchment, on the other hand, possess considerable affinity for the coal-tar colours. The same is the case with *jute*, which consists essentially of *bastose*, a substance closely allied to, but not identical with, cellulose. *Chinagrass* also evinces a certain affinity for the basic coal-tar colours.

It is at present almost generally admitted that the combinations of the substantive colouring matters with the fibres are of a chemical nature. In certain cases, however, it is difficult to conceive that a chemical combination takes place, and in these it seems more rational to ascribe the dyeing to a mechanical adhesion or attraction than to a chemical reaction. Thus, alkali-blue, which is the sodium salt of the mono-sulphonic acid of aniline-blue, is absorbed from its neutral, colourless solution by wool or silk; but the actual colour, the free sulphonic acid, is only produced by a subsequent development in an acid bath. In this case, taking the chemical theory as the correct one, a chemical combination must have taken place between the fibre and the colourless sodium salt, a reaction which it is not easy to conceive from a chemical point of view. Another striking example is furnished by the *indigo vat*. If cotton is immersed for some time in an indigo vat, it will withdraw from the vat a much larger proportion of indigo-white than would correspond to the amount of liquid absorbed. A precipitation has therefore taken place in or on the fibre, but at the same time it cannot be supposed that a chemical combination has taken place between the indigo-white and a totally neutral and indifferent substance like cotton.

The substantive colours when dyed on the fibre possess, as a rule, the same colour as their solution in transmitted light. The adjective colours are, on the other hand, apt to vary considerably in shade, according to the mordant or other substances used in preparing the material. In silk or wool-dyeing, these variations in shades are sometimes brought about by first mordanting the material with a substantive colouring matter, and then dyeing with an adjective one.

Another method of fixing the substantive colouring matters is often resorted to in the printing of silk fabrics. The pieces are first charged with stannic oxide or alumina, by passing them through dilute solutions of stannate of soda or basic aluminium sulphate, after which they are dyed and steamed. The colours obtained in this way are much brighter than those obtained on material which has not been previously prepared. If, however, the silk is charged with a large amount of tin, the light shades are apt to be spoiled. This is the case with silk which has been weighted by being immersed repeatedly in a concentrated solution of stannic chloride and then washed in running water.*

ADJECTIVE DYEING.—There are, with the exception of the so-called *benzidine* colours, only a few of the coal-tar colours which possess a direct affinity for cotton, and the shades obtained with them are not permanent. Before dyeing, cotton must therefore be mordanted ; *i.e.*, it must be charged with some substance or substances which caus it to take up the colour. The mordants which are use. for the aniline dyes can be divided into two classes.

One of these yields insoluble compounds with the colouring matters which are precipitated on the fibre in definite quantities corresponding to chemical equations; as an instance of this class, we may mention the metallic hydrates which are used as mordants.

The other class of mordants does not yield de chemical compounds with the colouring matters simply absorbs the dye substantively. Thus, alb belongs to this class, as well as a large number of substances which are chemically quite indifferent i colouring matters, and only assimilate them on accou their physical structure, just as animal charcoal act   n coloured solutions. Many finely-divided precipitates,   ch as calcium phosphate, carbonate of lime, silicic acid, starch, etc., produced on the fibre, possess this property.

Metallic hydrates.—The principal mordants belonging to this category are salts of chromium, iron and alumina, tin and lead. They are used chiefly for the fixation of the acid colouring matters, such as eosin, cœruleïn, alizarin,

* Silk is sometimes weighted as much as 25 per cent. in this manner.

etc., with which they yield insoluble salts or lakes. These mordants are generally used in dyeing in the shape of soluble salts, such as lead acetate, stannic chloride, ferrous acetate, aluminium acetate, etc.

The usual process of mordanting cotton yarn with mordants like lead or alumina may be explained in the following manner:—If soda is added to a solution of lead acetate or aluminium sulphate in small quantities at a time, it will be noticed that these additions can be con- inued for some time without the appearance of a pre- ipitate, soluble basic salts of lead, or alumina, being formed.

If cotton yarn is now immersed in a solution prepared in this manner, these basic salts are decomposed even in the cold, but much more rapidly on the application of heat, into more basic, insoluble ones, which are fixed by the fibre, while at the same time acid salts are formed which remain in solution. The fibre possesses, therefore, an affinity for the mordant, which is most likely due to a dialytic action of the former.

The complete fixation of the mordants is effected by ing the material after mordanting through dilute tions of soap, soda, or chalk, or by washing in calca- us water.

The affinity of the woollen fibre for mordants is much ater than that of cotton. For if wool is heated in a tion of alum, the assimilation of basic salts of alumina begins at once. An addition of tartar to the solution of the mordant is beneficial, since it not only renders the deposition of alumina on the fibre more copious, but more even. In the case of *bichromate of potash* or bichromate of soda, which are used largely in mordanting wool for the alizarin colours, the action appears to be essentially of a strictly chemical nature, the chromic acid combining with some constituent of the fibre to form an insoluble or sparingly soluble chromate, while neutral chromate re-

E

mains in solution (See *Journ. Soc. Dyers and Col.*, 1887, p. 118).

The chromium precipitated in this manner on the fibre is consequently still in the form of chromic acid, or chromate, and is in this form not suitable for the fixation of certain colouring matters, like galleïn, alizarin-blue, etc. For the purpose of obviating this, and in order to obtain the chromium on the fibre in the green state—*i.e.*, as chromic hydrate—tartar is used along with the bichromate in mordanting; the tartar is oxidized at the expense of the chromic acid on the fibre, which is reduced to chromic hydrate. In place of tartar, tartaric acid or oxalic acid may be used. "Green mordanted" wool may also be obtained by using chromium fluoride (about 6 per cent.) in place of bichromate.

As a rule, mixed mordants yield more stable colours than single ones, and experience has shown that it is most advantageous to use a sesqui-oxide (ferric oxide, chromic oxide, alumina) along with a monoxide (lime, magnesia, stannous hydrate, zinc hydrate). The double colour-lakes subsequently produced in dyeing are characterized by their fastness to acids and alkalies; an admixture of magnesia makes the alumina-lakes fast to alkalies.

The application of these mordants in printing will be referred to again under alizarin.

Sulphur.—The insoluble modification of sulphur has no affinity whatever to the colouring matters, but the amorphous modification has a very peculiar action. If wool is impregnated with a solution of sodium hyposulphite (sodium thiosulphate) and is then passed through a dilute mineral acid, such as sulphuric acid, amorphous sulphur is precipitated on the fibre, according to the following equation .—

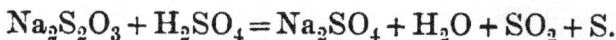

$$Na_2S_2O_3 + H_2SO_4 = Na_2SO_4 + H_2O + SO_2 + S.$$

The fibre is thus rendered capable of absorbing certain

colouring matters much better than before this treatment. This is especially the case with methyl-green, with which it is difficult to obtain bright and even colours on unmordanted wool; malachite green is also improved.

Metallic sulphides.—The sulphides of zinc and tin precipitated on the cotton fibre by double decomposition render the latter capable of being dyed with magenta, methyl-violet, and Bismarck-brown. Very full shades can be obtained in this manner, which are at the same time pretty fast to washing. Aniline-blue and safranine do not give satisfactory results by this process.

Silica.—Finely-divided or gelatinous silica readily absorbs the aniline dyes. Although it has not yet been positively proved, it is nevertheless probable that the combinations obtained in this manner are not of a chemical nature, but purely mechanical, and are due to the structure of the finely-divided silica. This property of silica is sometimes made use of in dyeing with alkali-blue. Soluble glass is added to the dye-bath when, in the subsequent development with acid, hydrated silica is separated out on the fibre, thus rendering the colour both faster and fuller.

Double cyanides.—The insoluble salts of hydroferrocyanic and hydroferricyanic acids possess an eminent affinity for the basic aniline dyes. Thus, cotton first passed through a solution of ferrocyanide or ferricyanide of potash, and then through zinc sulphate, can be dyed with magenta, methyl-violet, methylene-blue, etc.

Oil mordants.—Several methods of dyeing cotton are based on the relations of the colour-bases to the fatty acids and soaps described on p. 39, all of which effect in one way or another the formation of the insoluble fatty acid compounds on the fibre. Acid colouring matters, like the eosines and the azo dyes, can also be fixed on the vegetable fibres previously prepared with fatty acids, the latter being dyed substantively.

One of the most usual oil mordants is the so-called

Turkey-red Oil, which is prepared by adding castor oil
or olive oil very gradually to concentrated sulphuric acid
(D. O. V.), care being taken to avoid any rise in temperature. The liquid is then neutralized with soda-ash, and
ammonia is added until a sample dissolves completely in
water. The whole is then allowed to stand for about
twelve hours, in order to allow the sodium sulphate to
crystallize out when the clear oil is decanted.

The oil mordants are seldom used alone, but generally
along with inorganic mordants, which cause them to
combine more intimately with the fibre, while at the same
time the colour becomes much faster. As an example of
the use of oil mordants in dyeing, the following process,
used in the manufacture of *half silk fabrics*, may be
described here.

Cheap fabrics consisting of silk and cotton are dyed in
the piece, because the silk used in their manufacture is of
so poor a quality that it will not stand the weaving operations if discharged in the yarn. It is therefore necessary
to prepare the cotton before weaving in such a manner as
to render it capable of absorbing the dye in exactly the
same proportion as the silk does. For this purpose the
cotton yarn is first impregnated with Turkey-red oil and
then" aged—" *i.e.*, exposed for some time to the action of
air and light—by which latter process the fatty acids are
liberated and undergo a peculiar partial oxidation, which
renders them insoluble in weak soda solutions. The yarn
is then passed through a bath of neutralized alum. After
having been prepared in this manner, it is made up into
fabrics with the raw silk, which is discharged in the piece,
and the fabric is finally dyed with magenta, eosin, aniline-
blue, scarlet, etc. Another way of making up this kind
of goods is first to weave the material, then discharge,
mordant, oil, and steam, in order to fix the oil, pass
through weak soda solution (to remove the surplus of
oil), and dye.

The colours produced with the basic coal-tar colours on cotton mordanted with Turkey-red oil are very brilliant, but are not at all fast to light.

Oil mordants are also used for colouring matters which can be fixed with inorganic mordants alone, e.g., for cœruleïn and alizarin ; the shades obtained are much faster and more brilliant than those obtained without the use of oil mordants.

In yarn-dyeing, it is not always necessary to make use of Turkey-red oil when an oil mordant is required. A very good substitute can be prepared by gradually mixing 1 kilo. of olive oil with 50 c.c. of concentrated sulphuric acid. The milky liquid thus obtained is added to the lukewarm bath, in which the yarns are worked until they are sufficiently impregnated. They are then wrung and transferred to the dye-bath, which contains, besides the dye, a small quantity of alum and soda.

Soap has also been proposed as a mordant for the aniline dyes. The cotton is first passed through a soap solution, dried and dyed without being previously washed. By the double decomposition of the soap and the colouring matter, insoluble compounds of the fatty acids and colour-bases are formed on the fibre. The colours obtained by this method are not stable.

Soap is also sometimes employed for the precipitation of the compounds of the higher fatty acids (olive, palmitic, and stearic acids), with alumina or lead on the vegetable fibres, in mordanting for such colours as spirit-blue and the eosins. The method usually adopted is to work the material first in a strong soap bath, wring it out well, and then pass it, without washing, into a bath of alum or lead acetate.

Sometimes fatty acids are introduced into the colouring matters after the latter have been fixed on the fibres by inorganic mordants, in order to produce brighter and faster colours. This is effected by passing the goods, after dyeing, through a boiling soap solution, or by oiling them with Turkey-red oil, or by other similar methods.

The important part which the oil mordants play in *alizarin-dyeing* will be referred to later on under that heading.

Tannin mordants (astringents).—Tannic acid will combine with the basic aniline dyes to form either soluble compounds or insoluble lakes, according to the proportions of tannic acid and colouring matter used.

According to the experiments of Juste Koechlin, insoluble lakes are obtained when the following proportions are made use of :—

4 pts. Magenta	5 pts. tannic acid	2 pts. soda crystals.
4 ,, Malachite-green	5 ,, ,, ,,	1 ,, ,, ,,
4 ,, Parma-violet	5 ,, ,, ,,	1 ,, ,, ,,
4 ,, Methyl-green	10 ,, ,, ,,	4 ,, ,, ,,

If cotton is steeped for some time in a solution of tannic acid, it will absorb a certain proportion of the latter, and is thus rendered capable of fixing the basic aniline dyes. The affinity of cotton for tannin is not, however, a very strong one, since by long-continued washing the latter can all be removed again from the fibre. Cotton is seldom mordanted with tannin alone, since the colours obtained in this way are not stable. They are soluble in excess of tannic acid. The tannin is usually fixed on the fibre in an insoluble form, by passing it after the tannin bath through a solution of some metallic salt or other substance capable of yielding an insoluble compound with tannin (*e.g.*, gelatine).

Tartar emetic* or stannic chloride (tin spirits) are most

* Within the last few years a number of substitutes for tartar emetic have appeared in the market, for nearly all of which it is claimed by the inventors that they are cheaper and give equally good or better results than tartar emetic. Without discussing their relative merits, the following may be mentioned here:—The double oxalate of antimony and potash, the double fluoride of antimony and soda, lactate of antimony, and an alkaline solution of antimony in glycerin.

generally used for this purpose, but in many cases it is more advantageous to use other metallic salts, such as ferrous acetate, ferric sulphate, zinc acetate, neutralized alum, lead acetate, etc. In working on the large scale, it is first necessary to find out experimentally the most suitable proportions between the tannic acid and the metallic mordant. Thus it has been found that 5 pts. of tannic acid require for their complete precipitation 1 pt. of tartar emetic and 1 pt. of soda crystals.

The choice of the inorganic mordant to be used along with the tannin depends on the nature of the colouring matter, and on the shade required, since one and the same colouring matter can give very different shades with different mordants. In dyeing cotton prepared in this manner, a treble compound or lake is obtained, consisting of tannin, metallic oxide, and colouring matter.

The compounds of tannic acid with the aniline dyes are soluble in methylated spirits and in acetic acid. If the solutions obtained in this way are thickened with starch, gum, etc., and then printed on mordanted fabrics, insoluble compounds of colouring matter, tannin, and mordant are obtained on the fibre in the subsequent operation of steaming. This process is used largely in calico-printing. By subsequently passing the material through a bath of tartar emetic the colours are rendered faster.

Weighting of silk with tannin.—The coal-tar colours seldom lose any of their brilliancy in being combined with tannin substances, providing the latter are not coloured in themselves. It is thus possible to weight silks dyed in bright colours, in all but very light shades, from 12 to 15 per cent., by steeping them for some time in cold solutions of pure tannin. The silk combines directly with the tannin, and though materially increased in weight, it does not part with any of its valuable properties by this treatment.

Albuminous and gelatinous substances.—The coal-tar colours

behave similarly with substances like albumen and gelatine as they do with silk and wool, which are indeed
chemically closely allied to the former. Vegetable fibres
which have been " animalised—" *i.e.*, coated with a thin
layer of some albuminous substance—can therefore be
dyed like wool or silk.

This property of the coal-tar colours is very seldom made
use of in dyeing, but it is used on a large scale in calico-
printing. For this purpose a concentrated solution of *egg
albumen*, or, for dark shades, *blood albumen*, is mixed with
the colouring matter, printed, and steamed. In this way
the albumen is coagulated, and remains behind on the
fibre as an adhesive coloured coating. In place of albumen,
an ammoniacal solution of *casein*, or alkaline, or slightly
acid solutions of gluten can be used, but the results
obtained are not so good.

Solutions of glue containing the aniline dyes, along with
a small percentage of bichromate of potash, can also be
used for printer's colours. The printed fabric is exposed
to the action of the light, which renders the glue insoluble.

Colouring matters as mordants.—In some cases the
colouring matters themselves which have been fixed on
the fibre can act as mordants. Thus, aniline-violet, which
of itself has very little affinity for the cotton fibre, can be
dyed very well on fast violet (ferric alizarate).

The fact that colours obtained with two substantive
colouring matters are often much faster than either
colouring matter by itself is most likely also due to a
similar cause.

Most of the benzidine colours act, when dyed on cotton,
as excellent mordants for the basic coal-tar colours. Thus
cotton dyed with chrysamine readily absorbs magenta,
safranine, malachite green, etc., from their cold or tepid
dilute aqueous solutions producing compound shades of
great brilliancy. For the production of bright shades
in this manner, it is essential in the case of chrysamine,

at least, that the temperature of the bath containing the basic dye should not exceed 60° or 70°.

Fast and loose colours.

The fastness of colours produced on the fibre not only depends on the stability of the colouring matter itself against external influences, but also on the stability of the combination between fibre and colouring matter. Thus a very fast colouring matter, such as Guignet's green, may be combined with the fibre in so loose a manner (by means of albumen) that it falls off in a powder in washing or wearing, without undergoing any material change itself.

A material is usually called fast-dyed if it will withstand every action to which it is likely to be exposed in the course of the use for which it was intended. Cotton fabrics, for example, ought to stand the process of washing much better than silk, since the latter is generally cleansed in quite a different way; while the Turkish fez, which is exposed all day to the action of the sun's rays on the head of the dockyard workman, must naturally be much faster to light than the costly silk dress, which is carefully stowed away in a dark wardrobe and but seldom sees the daylight.

General regulations as to the experiments which should be carried out in order to test the fastness of a colour can therefore not be of much use, and in every case the judgment should be guided by circumstances.

A material is called *fast to washing* if it will stand boiling with a neutral or slightly alkaline soap, without changing or losing any appreciable quantity of its colour. Some colours will stand boiling with dilute solutions of caustic alkalies, and are then called fast to alkalies; they are very valuable on cotton or linen goods.

In the cloth manufacture it is of great importance that the colours should be *fast to milling*. The term " milling " embraces all those operations which are calculated to effect the felting of the woollen fibres in the fabric by means of pressure or friction, along with fuller's earth and solutions of soap, soda, stale urine, etc. The better qualities of cloth are dyed in the yarn, or sometimes in the loose wool. From this it is evident that the colours must necessarily withstand the subsequent treatment with alkaline liquids.

Some colours can be rendered faster in milling by the addition of certain substances to the dye-bath. Thus, an addition of magnesium sulphate has the effect of neutralizing the effect of the liquid used in milling by the separation of indifferent magnesium hydrate.

Every fabric should be sufficiently *fast to acids* to withstand the action of perspiration, and should therefore not be changed by the organic acids contained in the latter Most of the dyes used for silk must of necessity possess this property, in order to stand the brightening process to which silk yarns or fabrics are almost invariably subjected after dyeing. *Sulphurous acid* has a peculiar bleaching action on many colouring matters. It is therefore a frequent occurrence that dyed goods which are stored in places where gas is burnt, fade in colour.

Heat, also, has an injurious action on many colouring matters, and this is especially seen in steaming silk or woollen fabrics after printing. The operation of steaming is carried out in the following manner:—The fabrics, after having been printed, are exposed in air-tight boxes or cylinders to the action of low-pressure steam, the temperature of which is not much above 100° C. Colours which will bear this treatment are called *fast to steaming*. Some colours begin to sublimate at this temperature, and are partially deposited on the white parts of the fabric, while others undergo a chemical decomposition. Thus, methyl-green, when steamed for some time, is transformed into methyl-violet. The blue obtained from propiolic acid is also affected in steaming. Some colouring matters are so volatile that fabrics dyed with them colour the paper in which they are wrapped.

Nearly all organic colouring matters are bleached by the continued action of *light* and *air*. Light alone is able to cause chemical changes, but at the same time it favours the formation of small quantities of ozone and peroxide of hydrogen, especially when water gradually evaporates from the surface of the fibre. But even if we do not take into consideration that fabrics become wet with rain, etc., this gradual evaporation and absorption is an almost continuous process, since the degree of moisture of the fibre is dependent upon the temperature and moisture of the surrounding atmosphere. Ozone and peroxide of hydrogen belong to the most powerful bleaching agents known, and exercise, therefore, a destructive action on the colouring matter.

The chemical effects which the different-coloured rays are able to produce vary greatly,—the red, yellow and green rays having little or no effect, while the blue, violet and ultra-violet rays possess the most powerful chemical action. For this reason, materials which are worn in gas-light or candle-light do not fade as rapidly as those which are exposed to daylight; for the light emitted by these artificial illuminating agents contains a much smaller proportion of blue and violet rays than daylight.

It can easily be ascertained whether a colour is *fast to light* by exposing one-half of the material to the action of direct sunlight, while the other is covered up. Colours which are not fast to light will sometimes show a marked change between the two halves within an hour or two, but certainly within twenty-four hours of direct sunlight. In carrying out the above experiment, it is advisable to expose simultaneously some other colour, which is considered sufficiently fast to light, since nearly all colouring matters lose some of their brilliancy by the action of light.

In place of direct sunlight, which is not always at hand, a powerful electric arc light may be used; or, failing this, diffused daylight may be collected and made to fall upon the material by means of a large lens. One and the same colouring matter may vary in fastness according to the material on which it is fixed. Thus, vat-indigo, when dyed on wool, fades much more rapidly than on silk or cotton; a fact which is most likely due to a reduction of the colouring matter in the interior of the fibre.

The Testing of Colouring Matters.

The complete analysis of a colouring matter is, generally speaking, one of the most difficult subjects which can be placed before a chemist. For it is not only necessary to determine the actual percentage of pure colouring matter, but also that of the impurities which are generally formed in the manufacture, and often so closely resemble the pure colouring matter in their chemical properties that their separation and estimation is rendered extremely difficult. The results of a complete analysis of a colouring matter do not, however, always give an exact criterion of the quality, since the presence of very small quantities of certain impurities may alter the shades considerably in dyeing; while, on the other hand, the presence of other impurities may not have any injurious effect on the shade.

Comparative dye-trials.

In order to obtain a rapid and reliable estimate of the value of a colouring matter, the best method is to dye with it on the small scale, using for the purpose the same material for which the colouring matter is intended on the large scale.

In carrying out these comparative dye-trials, hanks of yarn of pieces of cloth are generally used, which are all wound or cut to a certain weight, which varies for wool and cotton from 5 to 20 grm., and for silk 2 to 5 grm.

The dye-trials are either carried out in beakers (glass or porcelain) or in small vessels of tin or tinned copper, having about the shape indicated in Fig. 13. They are much higher than broad, and the rim is provided with two indentures calculated to hold the glass rod on which the yarn is suspended. Never less than two dye-trials

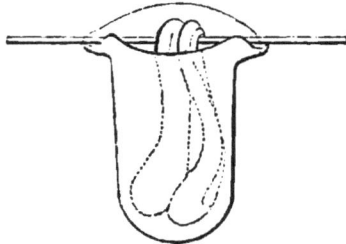

FIG. 13.

should be carried out at once, viz., one with the new colouring matter, the other with a colouring matter of known value, which is taken as the standard. It is, however, frequently necessary to compare a larger number of samples with the standard.

It is absolutely necessary that all trials should be carried out under exactly the same conditions as to temperature, time of immersion, etc. In order to effect this, it is necessary in the first place to have the dye vessels all of the same material, as uniform as possible in thickness, the quantity of liquid the same in all cases ; and in order, lastly, to ensure that the temperature is the same in each vessel, they should all be placed in a water-bath, or if a higher temperature is necessary, in a glycerin or oil-bath.

The following example will serve to illustrate the method of testing a new colouring matter in the manner described above. Two vessels (glass or tinned copper) are chosen of equal size, and into each is placed the same amount (200 to 600 c.c.) of lukewarm water, as well as equal amounts of those additions which are necessary in using the dye on the large scale, such as sulphuric acid, soap, boiled-off liquor, etc.   The weighed hank of yarn or " swatch " of cloth to be used in the experiment is then thoroughly wetted, and immersed in the liquid.

Solutions of known strength of the two colouring matters to be compared are then made by dissolving accurately weighed quantities (from 0·1 to 1 grm.) in 100 c.c. of water or methylated spirits.

The hanks or swatches are then taken out, and equal volumes of each colour solution are added to the respective dye vessels, and after stirring the contents well, the material is again introduced and dyed with a gradual rise of temperature.   When the baths are exhausted, further quantities of the colour solutions are added gradually, until both patterns have acquired the proper degree of saturation and appear equal in shade.   In working in this manner, it is often found that unequal quantities of the colouring matters are necessary to produce the same shade; and if the quantities added have been accurately measured as described above, this method will not only serve as a comparative test for the purity of shade, but will also give an idea as to the quantitative value of the colouring matter.   If, for instance, 9 c.c. of the solution of the colouring matter taken as type had given the same depth of shade as 13 c.c. of the solution of the new colouring matter, we should infer that 100 pts. of the type were as strong as 144 pts. of the sample.

Although it is possible to compare the shades of the two samples while they are being dyed, it is nevertheless advisable to make the final comparison only after they have been washed and dried.   With some practice, it can easily be seen whether two colours are identical, or whether one of them appears a little less pure or duller.   Even very slight differences in the tone of the colour, which an outsider can scarcely distinguish, are readily perceived by a practical man. Artificial light is a very important item in comparing shades, especially for the green, blue, and violet, since by this means certain peculiarities are brought out very distinctly. Thus, a blue with a slight cast of red appears almost violet in artificial light (gas or lamp light), while a blue with a slight shade of green appears distinctly bluish-green.

It should still be mentioned that, if possible (see below) the dye-bath should be completely exhausted, i.e., the whole of the colouring matter should be taken up by the fibre, and the bath should appear colourless or nearly so. Otherwise erroneous results may be obtained, since the impurities which have the power of dyeing are generally fixed last by the fibre. The presence of impurities of this kind can easily be detected by preparing a solution of the colouring matter and then dyeing two hanks in it, one after the other, so that the second completely exhausts the bath. The second hank may then be compared with the first, or with another dyed to the same depth of shade in a fresh solution.

Many colouring matters, such as picric acid, etc., cannot be completely withdrawn from their solutions in dyeing. In these cases, equal weights of the colouring matters in question are used for the dye-trials, and the colours produced are compared directly with respect to saturation and purity.

In testing adjective colouring matters, mordanted yarn or cloth must be used, and, after dyeing, the material must be made to pass through all those operations (soaping, clearing, etc.) which it has to undergo on the large scale. Since the treatment of the different adjective colouring matters varies considerably, no general instructions can be given here.

Lastly, in giving preference to one of several samples of a colouring matter, the price should be taken into account, besides the results of the dye-trials, since it may in many cases be more advantageous to make use of a weaker dye, in case the price is considerably lower. Besides, in dyeing dark or mixed shades, cheaper qualities of a colouring matter can often be used, which may not yield pure shades by themselves.

Colorimetry.

The estimation of the amount of pure colouring matter contained in a sample by means of a colorimeter does not give as reliable results as a comparative dye-trial, and has therefore but little technical importance. For this reason the principle of the colorimeter will only be referred to in a few words. Two glass tubes, closed at one end, and being of exactly the same diameter, are placed close to each other on a stand. Each tube is divided into a certain number of equal parts, say, from 0 to 200. Equal weights of the colouring matters to be compared are dissolved in water (alcohol, etc.), and the solutions are poured into the tubes up to the mark 100, when as a rule one will appear somewhat darker than the other. The darker liquid is then diluted with water in small quantities at a time, until both liquids, when looked at horizontally, appear to have the same strength. If, in order to effect this, 35 c.c. had been added to the normal solution (the solution of the type), we should conclude that the dyeing power of the type compared with that of the sample is as 135 : 100.

An ingenious instrument for the measurement of colour, either in solution or on dyed fabrics, is the *tintometer*, devised by Lovibond, for a full description of which, see *Journ. Soc. Dyers and Col.*, 1887, p. 186.

Quantitative determination of the coal-tar colours.

Until recently no ready and accurate methods were known by means of which the coal-tar colours could be quantitatively determined. In a communication to the

Society of Dyers and Colourists (*Journ. Soc. Dyers and Col.*, 1888, p. 82), however, Rawson describes a process for the valuation of naphthol yellow, by means of which the percentage of pure colouring matter can be accurately and rapidly determined in a very simple manner. The process, which is also applicable to picric acid, tartrazine, and most of the azo dyes, depends upon the complete precipitation of these dyes by a solution of night-blue * in acetic acid. The following is the *modus operandi* in the case of naphthol yellow or picric acid :—

A standard solution of night-blue is prepared by dissolving 10 grms. of the dye in 50 c.c. glacial acetic acid and diluting to 1 litre. Solutions of the samples of naphthol yellow or picric acid are then prepared so as to contain 1 grm. of the dye per litre. 10 c.c. of the night-blue solution are now carefully measured into a small flask, and then about 30 c.c. of the yellow run in from a burette. After shaking and allowing to stand for about a minute, the contents of the flask are poured through a filter into a colourless Nessler glass. If the filtrate is blue or colourless, more of the yellow solution is added until the filtrate shows a very faint yellow tint. The value of the samples under examination will be in inverse proportion to the number of cubic centimetres required to precipitate 10 c.c. of the night-blue solution. Thus if of two samples, 28 c.c. are required in the one case, and 35 c.c. in the other, their relative value will be as 35 : 28, or as 100 : 80.

In applying the method for the valuation of azo-dyes, it is necessary to use it with discretion, since colouring matters to be compared with each other must possess the same chemical constitution. It would not be admissible to compare, for example, a crocëin scarlet with a xylidine scarlet in this manner; but a number of samples of either of these scarlets, containing various proportions of either

---

* Tetramethyltolyltriamidodiphenylnaphthylcarbinol hydrochloride.

F

mineral or organic impurities, can be valued by the night-blue process with great precision.

The reverse process is applicable for the estimation of night-blue and of crystal violet.

Impurities in colouring matters.

The impurities found in the artificial colouring matters may either be accidental, and result from the mode of manufacture, or they may have been added for special . purposes. In the latter cases, however, this is seldom done for the purpose of adulteration. Thus, pastes are often mixed with glycerin, in order to prevent the solid constituents settling into a hard cake. On the other hand, many very powerful dyes would necessitate great care in weighing when used in small quantities, and are there-fore delivered, according to the wish of the consumer, in a state of dilution, which is effected by mixing them with a certain precentage of dextrin, sugar, salt, or some other harmless substance.

The substances used for the adulteration of colouring matters are usually chosen so as to fit the nature of the respective dyes, and are not easily detected in the ordinary application of the same. In many cases a mi-croscopical investigation renders good services. For the detection and quantitative estimation of the admixtures, it is, however, necessary to resort to special methods de-pending on the properties of the adulterated dye-ware, and it is therefore difficult to give any general instruc-tions here.

In all cases, the water and the ash of the sample should be estimated. The estimation of the water is best effected by heating a weighed quantity of the sample in a drying oven to about 160° C. Some colouring matters contain water of crystallization, which is also given off at this tempera-ture, and should be subtracted from the total, to obtain the actual *percentage of moisture*. In estimating the *dry substance of pastes*, it is not sufficient to take a weighed sample, dry and weigh it, since any other substances contained in the water, such as inorganic salts, glycerin, etc., would also be contained in the residue. Since pastes are often difficult to filter after having been diluted with water, the best method is to evaporate to dryness in a small mortar, after which the residue is finely pulverized, and treated with water. The residue, which has now assumed a pulverulent condition, is collected on a tared filter, washed out well with water, dried at 100° and weighed.

By the *estimation of the ash*, the presence of inorganic adulterants can easily be detected. It should, however, be borne in mind that certain salts may have become mixed with the dye in the ordinary process of manufacture, and that many colouring matters are combined with inorganic bases, or are brought into commerce in the shape of double salts, and must therefore of necessity contain a certain precentage of ash, which can easily be calculated from the formula.

Organic impurities can frequently be detected by dissolving a sample in water, and withdrawing the colouring matter from the solution by means of wool or silk. The impurities which have remained in solution will be contained in the residue on evaporation, and can then be subjected to a closer investigation. An admixture of *cane sugar* can easily be detected in the following manner, should the colouring matter be soluble in absolute alcohol:—The sample is extracted with absolute alcohol containing a little ether, until the residue of sugar is rendered almost colourless, when it is collected on a tared filter and weighed. In case the colouring matter is soluble in water and can be precipitated by acetate of lead, another method may be used. A weighed quantity of the sample is dissolved in water and treated with ex-

cess of acetate of lead, when the liquid is filtered, and the
sugar is estimated in the filtrate by means of the polariscope. For the estimation of cane sugar in magenta, a
colouring matter which is not precipitated by acetate of
lead, this method has been modified in the following
manner:—A weighed sample is dissolved in water, the
rosaniline is precipitated by picric acid, the picric acid by
acetate of lead, and the liquid thus freed from colouring
matter is analysed in the polariscope.

Dextrin, too, is often used for adulterating or diluting
the coal-tar colours, and in some cases gum-arabic is used
for this purpose. The detection of these substances depends upon their insolubility in alcohol. If the colouring matters are soluble in water, the following is the best
method to adopt:—

From 1 to 2 grams of the substance are dissolved in as
little water as possible, and the solution is filtered into a
beaker, which has been previously weighed, along with a
glass rod. On adding a sufficient quantity of alcohol, the
whole of the dextrin is precipitated in flakes, which on
stirring form a coherent mass, which adheres partly to the
glass rod and partly to the sides of the vessel. The liquid
is poured off, while the beaker and glass rod are washed
with absolute alcohol, dried at 110°, and weighed.

The quantitative estimation of dextrin in colouring
matters which are soluble in alcohol is effected by ex-
hausting repeatedly with strong alcohol, after which the
residue is dissolved in water and reprecipitated with alcohol in the manner described above.

PART II.

COAL-TAR AND THE RAW PRODUCTS USED IN THE MANUFACTURE OF THE COAL-TAR COLOURS.

If animal or vegetable substances are heated in vessels without access of air, a complete decomposition takes place. A large number of volatile products is formed, which are partially condensed in the receivers in a liquid or semi-solid state; another portion passes over in the form of gases, while the residue which remains in the retort consists chiefly of carbon. The condensed product usually consists of two distinct layers: an aqueous one, containing several substances in solution; and another one, which is generally of a dark colour, and is known as tar.

The products of the dry distillation of coal, which is carried out on the large scale for the manufacture of coal-gas, may be conveniently divided into four classes:— Coal-gas, ammoniacal liquor, coal-tar, and coke. Coal-tar is the only one of these which is of importance in the manufacture of the artificial colouring matters. It consists of a large number of different substances, which, according to their chemical reactions, can be divided into three groups.

The first of these comprises the *hydrocarbons*, which, as their name implies, consist of carbon and hydrogen. They are indifferent substances, possessing neither acid nor basic properties, and are therefore insoluble in dilute acids and alkalies. They constitute the principal part, and at the same time the most valuable part, of coal-tar. Benzene, toluene, xylene, naphthalene, and anthracene are the most important of these hydrocarbons.

The number of substances contained in the second group is limited, but they exceed those of the third group in quantity. They consist of the *phenols*, bodies which are composed of carbon, hydrogen, and oxygen. The phenols are weak acids, and therefore dissolve in caustic alkalies, whereas in dilute acids they are insoluble. The most important phenols contained in coal-tar are carbolic acid and cressol.

The third group, lastly, comprises a large number of *bases*, but none of these are contained in sufficiently large quantity to admit of their technical preparation from this source. The bases are composed of carbon, hydrogen, and nitrogen. They are soluble in acids, but insoluble in alkalies. As characteristic members of this group may be mentioned aniline, toluidine, etc.

Almost all those products which are contained in coal-tar in large quantities have been successfully utilised in the manufacture of colouring matters. The most important of these are benzene, toluene, xylene, naphthalene, anthracene, and phenol.

The separation and preparation of these products in the pure state forms a special branch of industry, *the distillation of coal-tar*. The portions of coal-tar which cannot be made use of in the preparation of colouring matters are generally used for other purposes. Thus one portion of the hydrocarbons is used as a solvent, under the name of " solvent naphtha "; crude carbolic acid is used for impregnating sleepers and for disinfecting purposes, while other portions of coal-tar are used as lubricants, and so on.

According to G. Schultz (Chemie des Steinkohlentheers),

the amount of coal-tar distilled annually in Europe is about 590,000 tons. This amount is subdivided as follows:—

England about 400,000 tons.
Germany   ,,   65,000 ,,
France   ,,   60,000 ,,
Belgium   ,,   50,000 ,,
Holland   ,,   15,000 ,,

The distillation of Coal-tar.

The separation of the different constituents of coal-tar is effected by means of fractional distillation, a process which depends upon the fact that, on heating a mixture of different liquids, the one which has the lowest boiling-point will pass over first into the distillate, the others following in order according to their boiling-points. As a simple example, we will take a mixture of absolute alcohol, which boils at 78° C., and water, which boils at 100° C. If this mixture is distilled, it will begin to boil at 78°, and at first almost pure alcohol will pass over. The separation by this ready means is, however, not an exact one, since very soon a part of the liquid of higher boiling-point is drawn over. If we were to divide the distillate obtained in our example into three equal parts, we should find that the first contained a strong alcohol, the second a very aqueous alcohol, while the third fraction would consist almost of pure water.

The boiling-points of the different constituents of coal-tar vary very considerably; benzene, for instance, boils at 80°, while anthracene boils at 370°. It is therefore possible in separating the fractions of the first distillation to obtain certain constituents in the first fraction only, while others are contained in the second or third.

The process of distillation is carried out in large iron retorts. The first distillate is divided into three portions, the first of which contains all the products which pass over

up to 180°, and which are technically known as *light oil*.
The second fraction consists of the *heavy oil*, so called because it sinks in water. The first portions of the heavy oil
which pass over remain liquid on cooling, but after a time
the distillate assumes the consistency of butter, owing to
the separation of solid naphthalene. The distillate then
becomes liquid again, and it is not until solid constituents
have begun to separate out again that the third fraction,
the so-called *green grease* or *anthracene oil*, is collected.
The residue which remains in the retort is known as *pitch*,
and is used in the preparation of asphalt, or of a fuel
known as " briquettes."

The light oil which passes over between 80° and 180° is
treated successively with soda ley, water, sulphuric acid,
and again with water, in order to free it from small
quantities of phenols and basic constituents, after which
it is distilled again.

The manufacture of pure benzene, toluene, and xylene
from the light oil purified in the manner described, is
carried out by fractional distillation. The stills used for
this purpose are provided with dephlegmators, by which
means it is possible to obtain almost pure products in one,
or at most in two operations.

The principle of these distilling apparatus is, not to
allow the vapour which is given off by the boiling liquid
to pass immediately into the condenser; but it is first
caused to pass through a series of pipes and vessels,
which are so arranged as to allow any liquid condensing
on their walls to flow back into the still. The temperature
of these vessels is regulated so as to allow only a portion
of the vapour, *i.e.*, that of the liquid of lowest boiling-
point, to pass through, while those possessing higher
boiling-points will be condensed and will flow back. If,
for instance, the liquid in the still consisted of a mixture
of benzene (B.P. 80°) and toluene (B.P. 111°), and the
vapour of the two were caused to pass through a series of

pipes heated to 80°, it is evident that the toluene will be condensed, while the benzene will pass through and can subsequently be liquefied in an ordinary condenser. It should not be imagined that by this process a quantitative separation can always be effected in one operation, although one form of condenser which is used on the large scale, and is known as " Savalle's column apparatus," will produce an almost pure benzene by the first distillation.

When the greater part of the benzene has been obtained, the temperature of the condensing vessels is raised, when a small quantity of liquid, consisting of a mixture of benzene and toluene, passes over and is collected separately. Afterwards pure toluene passes over for a length of time. Then comes xylene, which can also be isolated in a like manner; but this is not always done. The liquid remaining in the stills after distilling off the benzene and toluene is generally made use of for other purposes (solvent naphtha, etc.).

The heavy oil yields *carbolic acid* and *naphthalene*. The naphthalene is separated out in the solid state when the oil cools. It is pressed, treated alternately with caustic soda ley and sulphuric acid, after which it is purified by distillation or sublimation.

The separation of the crude carbolic acid, which consists chiefly of a mixture of phenol and cressol, is based upon the property of all phenols to dissolve in caustic alkalies. The heavy oil is agitated with caustic soda ley, and the aqueous layer having been drawn off, is exposed to the air for some time. This exposure causes certain impurities to separate out in the form of a brown resinous substance. The liquid is then filtered and the phenols are set free by the addition of an acid, when they are obtained in the form of an oil. Many methods are in vogue for the separation of pure phenol and cressol from this product, but the most rapid is a fractional distillation in a dephlegmation apparatus.

Commercial *anthracene* is obtained in the following manner from the so-called "green grease":—The oil is removed as much as possible by filtration, and the residue is pressed between warm plates, after which it is treated in a fine state of division with solvent naphtha. It is then pressed again, and finally sublimated with a current of steam. In this form the product is adapted for the manufacture of alizarin, although it still contains from 50 to 55 per cent. of impurities, which are very difficult to remove on the large scale.

The following is a list of the most important constituents of coal-tar :—

Constituents of Coal-tar.

A. *Neutral substances.*

Those printed in italics are commercially the most important.

Name.	Formula.	Melting point.	Boiling point.
Fatty hydrocarbons .	CnH_{2n+2}	—	—
,, ,, .	CnH_{2n}	—	—
Benzene	C_6H_6	+3°	81°
Toluene	C_7H_8	liquid	111°
Orthoxylene . . .	C_8H_{10}	,,	141°
Metaxylene . . .	,,	,,	141°
Paraxylene . . .	,,	15°	137°
Mesitylene . . .	C_9H_{12}	liquid	163°
Pseudocumene . . .	,,	,,	166°
Naphthalene . . .	$C_{10}H_8$	80°	217°
Diphenyl. . . .	$C_{12}H_{10}$	71°	254°
Fluorene	$C_{13}H_{10}$	113°	295°
Phenanthrene . . .	$C_{14}H_{10}$	100°	340°
Anthracene . . .	$C_{14}H_{10}$	213°	360°
Thiophene	C_4H_4S	liquid	84°
Carbazol	$C_{12}H_9N$	238°	355°

B. *Acid substances.*

Name.	Formula.	Melting point.	Boiling point.
Phenol	C_6H_6O	42°	182°
Orthocressol	C_7H_8O	31°	188°
Metacressol	C_7H_8O	liquid	201°
Paracressol	C_7H_8O	36°	199°
α Naphthol	$C_{10}H_8O$	94°	280°
β Naphthol	$C_{10}H_8O$	123°	286°

C. *Basic substances.*

Name.	Formula.	Melting point.	Boiling point.
Pyridine	C_5H_5N	liquid	115°
Lutidine	C_7H_9N	,,	154°
Aniline	C_6H_7N	,,	182°
Quinoline	C_9H_7N	,,	239°

COAL-TAR RAW PRODUCTS.

Constitution of the aromatic compounds.

According to the chemical theory at present generally adopted, the smallest particles of the elements which can enter into chemical combination are called *atoms.* These atoms combine with each other according to definite laws, in greater or smaller numbers, to form *molecules*, which are simultaneously the smallest particles of matter capable of existing in the free state. Thus each molecule of magenta is composed of atoms of carbon, hydrogen, nitrogen, and chlorine.

The abridged chemical method of specifying the number of these atoms and the proportion in which they are contained in the molecule constitutes what is known as an *empirical formula.* Thus the fact that the molecule of magenta consists of twenty atoms of carbon (C), twenty atoms of hydrogen (H), three atoms of nitrogen (N), and one atom of chlorine (Cl), is expressed by the empirical formula, $C_{20}H_{20}N_3Cl$.

By a thorough investigation of the chemical properties of these bodies, their modes of formation, products of decomposition, and the compounds they are capable of forming, modern chemistry has produced a deeper insight into their interior structure. But although we have no definite idea as to the actual relative position of the atoms in the molecule, we are nevertheless enabled to lay down a scheme in which the atoms given by empirical formulæ are arranged in such a manner as to show the whole chemical nature of the compound. A scheme of this kind, which simply serves to express our knowledge of the compound to which it relates, is called a *constitutional formula*. The determination of the constitutional formulæ of bodies like the organic colouring matters, the composition of which is generally very complicated, is often a difficult matter, and it is therefore not to be wondered at that there are still many colouring matters the constitution of which is unknown.

It is usual in chemical literature to deal with the elements and their compounds under two distinct headings; *viz., Inorganic Chemistry* and *Organic Chemistry*. The chief reason for this classification is that the organic compounds, or compounds of *carbon* with hydrogen, oxygen, nitrogen, etc., by far outnumber those of all the inorganic elements put together. The carbon compounds are again subdivided into— (a) The fatty series. (b) The aromatic or benzene series. (c) Pyridine and quinoline derivatives. (d) Bodies of unknown constitution. The coal-tar colours belong essentially to the second of these great classes; *i.e.,* they are derivatives of benzene and its homologues, of which benzene itself may be taken as the simplest representative. It contains six atoms of carbon, which are bound together (according to Kekulé's hypothesis, first expressed in 1865) by alternate double and single bonds in the form of a ring—

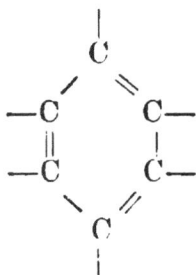

$$
\begin{array}{c}
\mid \\
\text{C} \\
\diagup \quad \diagdown \diagdown \\
-\text{C} \qquad \text{C}- \\
\parallel \qquad \mid \\
-\text{C} \qquad \text{C}- \\
\diagdown \quad \diagup\diagup \\
\text{C} \\
\mid
\end{array}
$$

Carbon being a tetrad or tetravalent element, it is obvious that each of these carbon atoms must still have one free bond. In benzene itself these bonds are saturated with hydrogen, and we have the empirical formula, C_6H_6, or the constitutional formula—

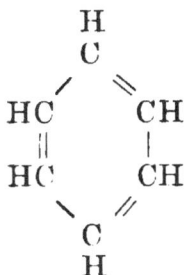

$$
\begin{array}{c}
\text{H} \\
\text{C} \\
\diagup \quad \diagdown\diagdown \\
\text{HC} \qquad \text{CH} \\
\parallel \qquad \mid \\
\text{HC} \qquad \text{CH} \\
\diagdown \quad \diagup\diagup \\
\text{C} \\
\text{H}
\end{array}
$$

By substituting for *one* of the hydrogen atoms in benzene other elements, or groups, such as Cl, NO_2 (nitro)-NH_2 (amido),OH (hydroxyl), etc., we obtain the so-called mono-substitution derivatives. The mono-derivatives are of one kind only; *i.e.*, there is only *one* nitrobenzene, $C_6H_5NO_2$, *one* chlorbenzene, $C_6H_5NO_2$, *one* amido-benzene, $C_6H_5NH_2$, etc.

When *two* of the hydrogen atoms in benzene are substituted by other elements or groups, the so-called disubstitution derivatives are obtained. Thus by treatment with chlorine we get dichlorbenzene, $C_6H_4Cl_2$, with nitric acid, dinitrobenzene, $C_6H_4(NO_2)_2$, and so on. But it is found in treating benzene with nitric acid, for instance, that not one, but three bodies are formed, each of which possesses different properties to the other two, while at the same time analysis shows them all to have the empirical formula, $C_6H_4N_2O_4$. It is here that Kekulé's benzene theory comes to our aid, and explains clearly why these three dinitrobenzenes should show different pro-perties. As we have already seen, benzene contains a ring of six atoms of carbon. These carbon atoms are usually numbered for convenience thus:—

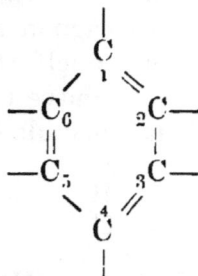

If now, in our example of dinitrobenzene, one NO_2 group attaches itself at 1, the other at 2, we shall get a dinitrobenzene of the formula—

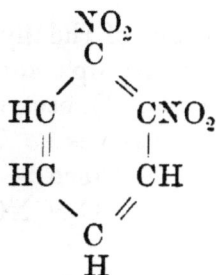

The substitution has taken place in two adjoining carbon atoms, and all such derivatives are known as *ortho-compounds*. The same compound would be formed if we substituted, instead of at 1 and 2, at 2 and 3, or 3 and 4, etc.

If the nitro groups are substituted at 1 and 3 (or 2 and 4, or 3 and 5, etc.), we obtain what is known as a *meta-compound*—

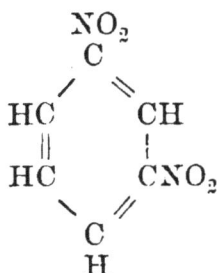

$$
\begin{array}{c}
NO_2 \\
C \\
HC \quad\quad CH \\
HC \quad\quad CNO_2 \\
C \\
H
\end{array}
$$

If, lastly, the substituting groups or elements lie opposite to each other in the benzene ring, the so-called *para-compounds* are produced. These may have the position 1 4, 2 5, or 3 6, and paradinitrobenzene would be represented thus :—

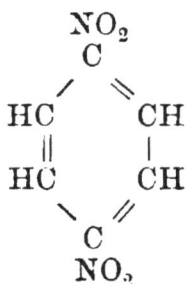

$$
\begin{array}{c}
NO_2 \\
C \\
HC \quad\quad CH \\
HC \quad\quad CH \\
C \\
NO_2
\end{array}
$$

Now what has been said of dinitrobenzene applies equally to all disubstitution derivatives of benzene. We invariably find three disubstitution derivatives, possessing the same empirical formula, but different constitutional formulæ. Such bodies are said to be *isomeric*, or are known as *isomers*.

When three elements, or groups of elements, are introduced into the benzene ring, the following cases of

isomerism are observed in case the substituting groups
are of the same kind (A) :—

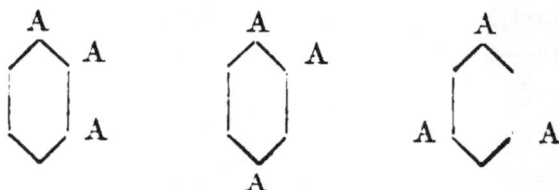

There are therefore *three* isomeric trisubstitutive derivatives.

Under similar conditions there are three tetra, one
penta, and one hexa substitutive derivative. When, however, in the tri, tetra, penta, and hexa derivatives the
substituting elements or groups are of different kinds, the
number of isomers is considerably increased.

The Hydrocarbons.

These bodies may be looked upon as the first raw
materials for the manufacture of the coal-tar colours.
They are obtained exclusively from coal-tar.

Benzene, C_6H_6.

Chemically pure benzene is a colourless mobile liquid,
possessing a specific gravity of 0·885 at 15° C. It boils at
80° C., and solidifies at 0° to a colourless mass, consisting of
rhombic crystals, which melt at +4°. It is almost insoluble in water, dilute acids, and alkalies, but it is easily
soluble in or mixable with alcohol, ether, chloroform, etc.
Benzene is an excellent solvent for fats, resin, etc., and
the impure product is frequently used on this account for
removing grease stains, etc. Benzene is easily acted upon
by concentrated sulphuric acid, nitric acid, and chlorine,

with formation of the corresponding substitution derivatives.

The purest form of *commercial* benzene is known as benzole for blue. Although it possesses exactly the same boiling point as pure benzene, it is not a chemically pure product, and contains, besides benzene, thiophene, and fatty hydrocarbons, which latter remain behind in the manufacture of nitrobenzene as " neutral oils." Besides benzole for blue, there are three other commercial products which are known respectively as 30 per cent., 50 per cent., and 90 per cent. benzole. The percentage indicates here the parts per hundred which pass over in the process of distillation under 100° C.

The valuation of commercial benzoles is effected by fractional distillation, and by determining the amount of nitrobenzene which they yield.

Toluene, $C_6H_5CH_3$.

Toluene may be regarded as the monomethyl derivative of benzene. It forms a colourless liquid, which boils at 111° C., and remains liquid at − 20°. At 15° it has a specific gravity of 0·872. With respect to solubility, etc., it possesses great similarity with benzene. By oxidizing agents it is converted into benzoic acid, C_6H_5COOH.

Toluene is contained in most commercial benzoles, in some of which it actually forms the chief constituent. Its presence in benzene for magenta is a matter of necessity. The pure product obtained by fractional distillation is used in the manufacture of malachite green and artificial indigo.

Xylenes, $C_6H_4(CH_3)_2$.

If two atoms of hydrogen in the benzene ring are replaced by methyl, three isomeric xylenes are formed which each possess the formula $C_6H_4(CH_3)_2$, but are dis-

G

tinguished from each other by their constitutional formulæ in the following manner :—

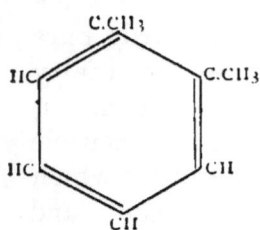 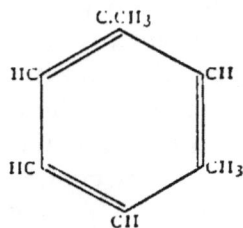 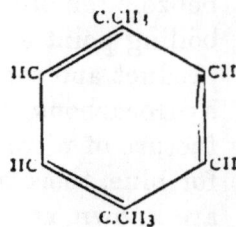

Orthoxylene.   Metaxylene.   Paraxylene.

That portion of coal-tar which passes over at about 140° consists of a mixture of the three xylenes, in which metaxylene invariably predominates. The isolation of the three isomers from commercial "xylole" cannot be effected by fractional distillation, since their boiling points lie so close together.

The preparation of metaxylene may be effected by boiling coal-tar xylole with dilute nitric acid for a length of time. The ortho and para xylenes thus become oxidized to the corresponding xylilic acids, while metaxylene is not acted upon. By a subsequent treatment with caustic soda the acids are removed and pure metaxylene is obtained.

Pure paraxylene can be obtained, according to Levinstein, by distilling commercial xylol with steam. The first portions of the distillate are cooled, and the crystals which separate are pressed and redistilled.

The separation of all the three isomers from xylole may be effected by Jacobsen's method. Xylol is shaken with strong sulphuric acid. Meta and orthoxylene are thereby converted into sulphonic acids, while paraxylene is not affected. The sulphonic acids are then converted into the corresponding sodium salts, which are separated from each other by crystallisation. The hydrocarbons are then regenerated by distilling the sodium salts of the sulphonic acids with ammonium chloride.

Orthoxylene is a colourless liquid, which boils at 141–142°.

Metaxylene, which forms about 70 per cent. of commercial xylole, is also a colourless liquid, has a specific gravity of 0·8668 at 19°, and boils at 139°.

Paraxylene forms colourless crystals, which melt at 15°, and boil at 138°.

Commercial xylole was formerly used as a solvent, under the name of "solvent naphtha," but since the discovery of the azo dyes it is used in large quantities for the production of xylidine and cumidine.

Hydrocarbons, $C_6H_3(CH_3)_3$.

Of these, mesitylene and pseudocumene are found in those portions of coal-tar which pass over between 140° and 180°. *Mesitylene* is symmetrical (1, 3, 5), trimethylbenzene, boiling point 163°; pseudocumene is asymmetrical (1, 3, 4), trimethylbenzene, boiling point 166°.

Stilbene, $C_{14}H_{12}$, or $C_6H_5CH:CHC_6H_5$.

This hydrocarbon has not hitherto been isolated from coal-tar. It is obtained by distilling sulphbenzaldehyde with finely divided copper. It forms colourless crystals, which melt at 125°, and boil at 307°. Diamidostilbene is used in the manufacture of one or two azo dyes which closely resemble the benzidine dyes.

Diphenyl, $C_{12}H_{10}$, or $C_6H_5-C_6H_5$.

Diphenyl may be regarded as phenylbenzene; *i.e.*, as benzene, in which one hydrogen atom has been replaced by phenyl (C_6H_5-). Although contained in coal-tar, it is not isolated, the quantity present being too small. It is best obtained by passing the vapours of benzene through red-hot tubes, and subjecting the product obtained to fractional distillation. Diphenyl forms colourless crystals, which melt at 71°. It boils at 254°.

Naphthalene, $C_{10}H_8$.

Naphthalene contains two benzene rings, and its constitution is shown by the formula—

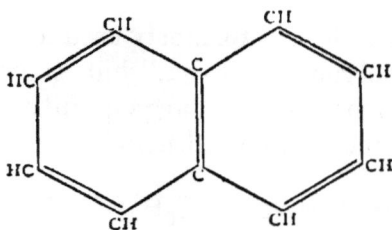

It will be seen, in looking at this formula, that in the formation of mono-substitution products two isomers are obtained, according as the hydrogen atoms marked β or those marked a are replaced in the following scheme:—

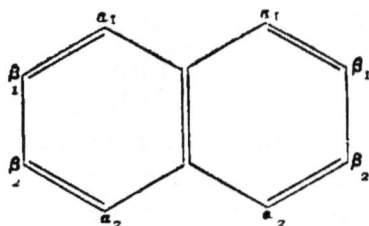

Thus, by replacing one hydrogen atom by OH two isomeric naphthols are obtained, which are known as alpha-naphthol and beta-naphthol respectively, but both possess the rational formula $C_{10}H_7OH$. For describing the monosubstitution derivatives, this designation would be sufficient, but in derivatives containing more than one substituting group it is necessary, in order to specify the constitution, to introduce the letters $a_1 a_2$ and $\beta_1 \beta_2$, as indicated in the diagram. When both substituting groups are contained in the same benzene nucleus, this is indicated in writing by joining the two letters with a single hyphen. Thus we should have for a dichlornaphthalene, in which the two chlorine atoms are in the same nucleus, and stand to each other in the orthoposition, the formula $C_{10}H_6Cl_2$ $(a_1 - \beta_1)$; if in the metaposition, $C_{10}H_6Cl_2$ $(a_1 - \beta_2)$; and if in the paraposition, $C_{10}H_6Cl_2$ $(a_1 - a_2)$.

When the substituting groups or elements are contained in different nuclei, this is represented by joining the letters with a double hyphen. Thus for a dichlornaphthalene, in which the two Cl atoms are not in the same nucleus, we should have the following possible cases of isomerism: $C_{10}H_6Cl_2$ $(a_1 = a_2)$, or $(a_1 = \beta_1)$, or $(a_1 = \beta_2)$, or $(a_1 = a_2)$, or $(\beta_1 = a_1)$, or $(\beta_1 = \beta_1)$, or $(\beta_1 = \beta_2)$, or $(\beta_1 = a_2)$, in all eight possible isomers.

The possible number of substitution derivatives in the naphthalene series, when the substituting elements or groups are of the same kind, will be,—

2 Monoderivatives.
10 Biderivatives.
14 Tri ,, ,,
22 Tetra ,,
14 Penta ,,
10 Hexa ,,
2 Hepta ,,
1 Octo ,,

When the substituting elements or groups are of different kinds the number of possible isomers becomes much larger.

Naphthalene was first discovered in coal-tar by Garden in 1820, but it was not until the introduction of the azo dyes that it became of much importance. The commercial product is obtained by the purification (crystallising, pressing, distilling, etc.) of that portion of coal-tar which passes over between 180° and 250°.

Pure naphthalene crystallises in leaflets or monoclinic tablets. It melts at 80°, and boils at 217°. The greater part of the naphthalene produced is used in the manufacture of the oxy-azo dyes.

Anthracene, $C_{12}H_{14}$.

The manufacture of this hydrocarbon has already been referred to in the distillation of coal-tar. It forms in the pure state colourless crystals, which show a beautiful blue fluorescence. Anthracene melts at 213°, and boils at a little above 360° C. Its constitutional formula is represented graphically in the following manner:—

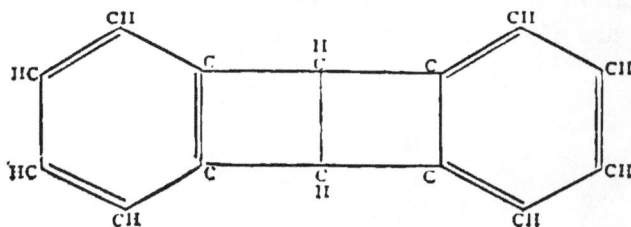

Anthracene is used almost exclusively for the manufacture of alizarin, anthrapurpurin, and flavopurpurin. The *valuation* of the commercial product is best effected by treating a known quantity with chromic acid in acetic acid solution. The crude anthraquinon thus formed is then treated with concentrated sulphuric acid at 100° C. for a few minutes, after which it is washed first with water, then with alkali, and then with water again. The pure anthraquinon is then dried at 100° C, and weighed,

Halogen derivatives.

By treating benzene and its homologues with chlorine, bromine, or iodine, substitution derivatives are obtained; *e.g.*,—

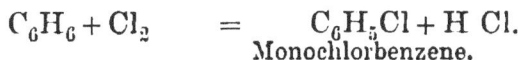

$$C_6H_6 + Cl_2 \quad = \quad C_6H_5Cl + H\ Cl.$$
Monochlorbenzene.

As a rule, chlorine acts most energetically (especially in presence of a trace of iodine), bromine less so, while iodine acts only very slowly. In the case of methylbenzene (toluene) and its homologues, the temperature plays a very important part in the formation of halogen substitution derivatives. If, for instance, chlorine is allowed to act on toluene in the cold, the hydrogen atoms of the benzene ring only are replaced, and we obtain $C_6H_4ClCH_3$, $C_6H_3Cl_2CH_3$, etc. If, on the other hand, chlorine is passed into boiling toluene, it is found that the hydrogen atoms of the methyl (CH_3) group only are replaced, and we get $C_6H_5CH_2Cl$, $C_6H_5CH\ Cl_2$, etc., bodies which differ entirely in their behaviour from those obtained in the cold.

The number of halogen derivatives in the aromatic series is great, but the following only are of commercial importance :—

Benzylchloride, $C_6H_5CH_2Cl$. This compound is obtained by passing chlorine into boiling toluene, until the latter has absorbed the theoretical quantity. It forms a colourless liquid, which has a specific gravity of 1·11, and which boils at 179°. It possesses a sharp, penetrating smell, and is converted by boiling with water (or better still, with a solution of K_2CO_3) into benzyl-alcohol, $C_6H_5CH_2OH$.

Benzalchloride, $C_6H_5CHCl_2$, is obtained like benzylchloride by the direct action of chlorine on boiling toluene. It forms a colourless, oily liquid, which has a specific gravity of 1·295, and which boils at 207°,

Benzotrichloride, $C_6H_5CCl_3$, is formed, like the two preceding compounds, by the direct action of chlorine on boiling toluene.   The crude product is purified by washing with water and dilute alkaline carbonates, after which it is dried with potash and distilled in a vacuum.   It forms a colourless, highly refractive liquid, which has a specific gravity of 1·38 (at 14°), and boils at 213°.

These products are used, either directly or indirectly, in the manufacture of malachite green.

Naphthalene tetrachloride, $C_{10}H_8Cl_4$, is a chlorine *addition* product of naphthalene, formed by the direct action of gaseous chlorine on molten naphthalene.   It forms colourless rhomboledric crystals, which melt at 182°. When treated with nitric acid, it is converted into *phthalic* and oxalic acids.

Nitro derivatives.

The nitro derivatives contain the group $-NO_2$ or $-N {\overset{=O}{\underset{=O}{}}}$.

They possess great importance, not only as intermediate products, but also in the form of colouring matters.   They are formed by the action of concentrated nitric acid upon the hydrocarbons and their derivatives.   Dilute nitric acid has only an oxidising action, and does not form nitro compounds.   If energetic nitration is required, fuming nitric acid (nitric acid containing lower oxides of nitrogen in solution) is frequently employed, either alone or in conjunction with concentrated sulphuric acid.   The object of the latter is simply to absorb the water formed in the reaction, which would otherwise dilute the nitric acid and render it less efficient.

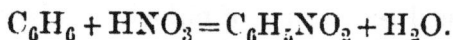

$$C_6H_6 + HNO_3 = C_6H_5NO_2 + H_2O.$$

In the manufacture of "nitrobenzole" on the large scale, sulphuric acid is almost invariably used for this reason.   Great care has to be taken to have the liquids thoroughly mixed, and to avoid (in most cases) any considerable rise in temperature.   In the manufacture of mononitro derivatives it is usual to allow the theoretical quantity of nitric acid (mixed with sulphuric acid) to flow into the hydrocarbon, so as always to have the latter in slight excess, and thus to avoid the formation of dinitro compounds.   In the case of di- and tri- nitro compounds, the reverse is the case ; *i.e.*, the hydrocarbon is allowed to flow into the mixture of the two acids.

The nitro compounds form either yellow or colourless oils or crystals, sparingly soluble in water.   Treated with alkaline reducing agents, they yield azo compounds, whereas by acid reducing agents they are converted into the corresponding amines.

Nitrobenzene, $C_6H_5NO_2$.   The manufacture of nitrobenzene (nitrobenzole) on the large scale is usually effected in large cast-iron cylinders of 200–400 litres capacity, which can be cooled from the outside by means of fine jets of water.   100 parts of purified benzene are (according to G. Schultz) introduced into the cylinder, and a mixture of 115 parts nitric and 160 parts sulphuric acids (both concentrated) is allowed to flow in gradually. At the beginning of the reaction it is necessary to cool, but towards the end (the reaction lasts from 8 to 10 hours) the temperature is allowed to rise gradually to 80°–90°. The contents of the cylinder are now run into a reservoir and allowed to settle.   Two layers are formed, the lower one consisting of the refuse acid (spec. gr. 1·63), the top one of nitrobenzene and unchanged benzene.*   The acid is drawn off, and the nitrobenzene, after having been washed with water, is freed from the unchanged benzene by treatment with a current of steam.   The benzene

---

* Besides unchanged benzene, this layer contains a certain proportion of hydrocarbons, possessing the same boiling point as benzene, which are not acted upon by nitric acid, and which appear to belong to the paraffin series,

which distills off along with a small quantity of nitro-benzene is used over again. The yield of nitrobenzene from pure benzene is about 150 per cent. (theoretical, 157 per cent.).

Nitrobenzene forms a yellowish, oily liquid, with a peculiar smell, similar to that of bitter almonds; it was, in fact, at one time sold as a perfume, under the name of *Essence de Mirbane*, but it is not used as such at present on account of its poisonous properties. It boils at 206–207°, and solidifies on cooling to needle-shaped crystals, which melt at +3°. Its specific gravity at 15° is 1·208. In water nitrobenzene is insoluble, but it dissolves easily in other ordinary solvents, such as alcohol, ether, etc., while at the same time it is itself an excellent solvent for many substances.

Commercial nitrobenzene is known either as *nitrobenzole for blue* or *nitrobenzole for red*. *Heavy nitrobenzole* contains very little nitrobenzene, and consists chiefly of nitro-toluenes and nitroxylenes. The first of these is used chiefly in the manufacture of aniline for blue and black and for induline, besides for the preparation of quinoline and benzidine. Nitrobenzole for red is used in the manu-facture of magenta.

Dinitrobenzenes, $C_6H_4(NO_2)_2$. According to theory, three isomeric dinitrobenzenes exist. They are all formed by the direct action of an excess of strong nitric acid on benzene or nitrobenzene. The meta compound forms the chief product of the reaction, while the ortho and para compounds are formed in comparatively small quantity only.

Metadinitrobenzene, $C_6H_4\begin{cases}NO_2 \ (1)\\NO_2 \ (3)\end{cases}$, is obtained on the large scale by allowing 100 pts. of benzene to flow into 100 pts. conc. nitric acid and 156 pts. conc. sulphuric acid. The nitrobenzene thus formed is then warmed for some time with a mixture of 100 pts. conc. nitric acid and

156 pts. conc. sulphuric acid. The product is poured into water (when it solidifies) and washed.

Pure metadinitrobenzene forms well-defined, almost colourless, needle-shaped crystals, which melt at 89·8°, and dissolvé in 17 pts. of alcohol at the ordinary temperature, while in boiling alcohol the compound dissolves very readily. It boils at 297°. Dinitrobenzene is used chiefly in the manufacture of phenylene diamine (for Bismarck brown).

Nitrotoluenes, $C_6H_4\begin{cases} CH_3 \\ NO_2 \end{cases}$. By the action of concentrated nitric acid on toluene the three isomeric nitrotoluenes (ortho, meta, and para) are formed. The ortho and para compounds are the chief products of the reaction, while of the meta compound a small quantity only is formed. The relative quantities of ortho and para nitrotoluene formed by the direct action of nitric acid depend to a great extent upon the concentration of the latter. The manufacture on the large scale is effected by allowing 10 pts. toluene to flow gradually into a mixture of 10 pts. conc. nitric acid and 15 pts. conc. sulphuric acid, contained in a vessel provided with a stirrer. The product of the reaction is washed and subjected to fractional distillation. In this manner the commercial ortho and para nitrotoluenes are obtained.

Orthonitrotoluene, $C_6H_4\begin{cases} CH_3\,(1) \\ NO_2\,(2) \end{cases}$, is liquid, and boils at 223°.

Metanitrotoluene, $C_6H_4\begin{cases} CH_3\,(1) \\ NO_2\,(3) \end{cases}$, melts at 16°, and boils at 230–231°.

Paranitrotoluene, $C_6H_4\begin{cases} CH_3\,(1) \\ NO_2\,(4) \end{cases}$, forms colourless prisms which melt at 54° and boil at 236°.

Dinitrotoluenes, $C_6H_3(CH_3)(NO_2)_2$. According to theory, four isomeric dinitrotoluenes exist, but of these the following only is of commercial importance:—

Alpha-dinitrotoluene, C_6H_3 $\begin{cases} CH_3\ (1) \\ NO_2\ (2). \\ NO_2\ (4) \end{cases}$ This compound is

formed by the action of nitric acid (in excess) on toluene. It forms long yellowish, needle-shaped crystals, which melt at 71°.

Nitroxylenes, $C_6H_3(CH_3)_2(NO_2)$.

According to theory, six isomeric nitroxylenes are possible. The most important (commercially) of these is the—

Alpha-nitrometaxylene, C_6H_3 $\begin{cases} CH_3\ (1) \\ CH_3\ (3). \\ NO_2\ (4) \end{cases}$ It forms the

chief product of the action of nitric acid or metaxylene; it boils at 238°.

Dinitrometaxylene—

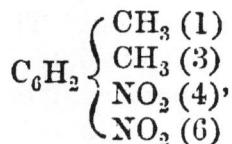

C_6H_2 $\begin{cases} CH_3\ (1) \\ CH_3\ (3) \\ NO_2\ (4), \\ NO_2\ (6) \end{cases}$

is obtained by the action of excess of nitric acid on metaxylene.

Nitronaphthalenes, $C_{10}H_2(NO_2)$.

According to theory, two isomeric mononitronaphthalenes are possible; viz., the alpha and the beta compound. The first of these is of commercial importance. It is obtained by the direct action of nitric acid on naphthalene, and forms fine yellow shining crystals, which melt at 61°. The boiling point is 304°.

Amido derivatives.

The aromatic amido derivatives, or *amines*, form the most important class of coal-tar raw products used in the manufacture of colouring matters. They may all be regarded as derivatives of ammonia—

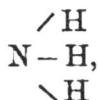

$$N \begin{cases} H \\ -H, \\ H \end{cases}$$

and are formed by replacing one, two, or all three of the hydrogen atoms in this compound by organic radicals. The amines are of three kinds. The *primary* amines contain two replaceable hydrogen atoms, the *secondary* amines one, while in the *tertiary* amines all the hydrogen atoms are replaced by organic radicals. The following may serve as examples :—

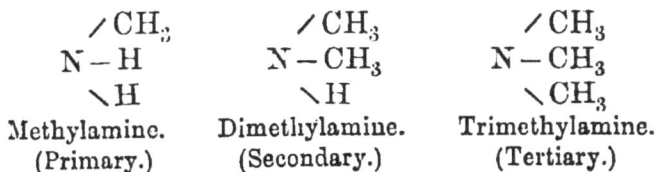

$N \begin{cases} CH_3 \\ -H \\ H \end{cases}$	$N \begin{cases} CH_3 \\ -CH_3 \\ H \end{cases}$	$N \begin{cases} CH_3 \\ -CH_3 \\ CH_3 \end{cases}$
Methylamine.	Dimethylamine.	Trimethylamine.
(Primary.)	(Secondary.)	(Tertiary.)

In place of methyl (CH_3) any other organic radical can be substituted; *e.g.*, ethyl (C_2H_5), phenyl (C_6H_5), benzyl (C_7H_7), etc.

In the amines, therefore, the nitrogen is always in direct combination with carbon and hydrogen, or with three carbon atoms. The group NH which is contained in the secondary amines is known as the Imido group.

The amido compounds are (with the exception of triphenylamine) of a basic character and combine with acids to form salts. The primary *diamines* contain two ·NH_2 groups, the *triamines*, three.

The aromatic amines are usually prepared by the reduction of the nitro compounds, in one or two instances also by reduction of the nitroso compounds by nascent hydrogen—

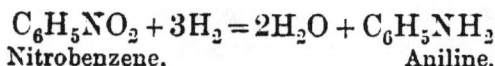

$$C_6H_5NO_2 + 3H_2 = 2H_2O + C_6H_5NH_2$$
Nitrobenzene.   Aniline.

This may be effected—

(1) With tin and hydrochloric acid.
(2) ,, iron and acetic acid.
(3) ,, ,, ,, hydrochloric acid.
(4) ,, sulphuretted hydrogen in ammoniacal alcoholic solution.

Of these methods, 3 and 4 only are used on the large scale, the first for the complete reduction, the last for the partial reduction, of the nitro compounds.

Amido compounds are also formed by the action of ammonia on the phenols at high temperatures—

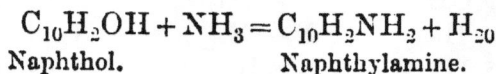

$$C_{10}H_2OH + NH_3 = C_{10}H_2NH_2 + H_2O$$
Naphthol.   Naphthylamine.

Primary amines.

Aniline, or phenylamine, $C_6H_5NH_2$, commercially known as "aniline oil," was first discovered by Unverdorben in 1826, who obtained it, among other substances, by the dry distillation of indigo. Aniline is a constituent of coal-tar, but occurs there in such small quantities that its isolation could not be made to pay.

The manufacture of aniline on the large scale is effected by reducing nitrobenzene with iron and hydrochloric acid. The operation is carried out in large cast-iron cylinders provided with stirrers. The materials are used in about the following proportions:—40 pts. water, 25 pts. iron borings, and 8–10 pts. hydrochloric acid, are placed in the apparatus, and 100 pts. nitrobenzene are allowed to flow

in, the stirrer being in motion. The contents of the
vessel are then heated by means of steam, in order to
start the reaction. The rest of the iron (85–100 pts.) is
then added gradually during eight hours. When all the
nitrobenzene has become converted into aniline, slaked
lime is added to the contents of the apparatus, in order to
liberate all the aniline, which is ultimately blown off by
a powerful current of steam.

Aniline forms a colourless oily liquid with a peculiar,
not disagreeable, smell. It boils at 182°, and has a specific
gravity at 15° of 1·0265. Aniline dissolves at the ordinary
temperature in 31 times its weight of water, and mixes
in all proportions with alcohol, ether, and other ordinary
solvents. When exposed to air and light, it soon turns
brown. Aniline possesses a burning taste, and acts when
taken internally as a powerful poison. By the action of
oxidising agents, aniline is transformed into various
colouring matters, many of which possess great commercial importance and will be referred to again.

Commercial "aniline oil" comes into commerce in four
different qualities, which are known as—

1. Aniline oil for blue and black.
2.   „   „   for red (magenta).
3.   „   „   safranine.
4. Liquid toluidine.

The first of these is almost chemically pure aniline. It
is used principally for aniline black and for the preparation of dimethylaniline, diphenylamine, quinoline,
phenylhydrazine, sulphanilic acid, and induline. Aniline
for black is usually tested by fractional distillation; 90
per cent. should pass over up to 182° C.

Aniline combines with the acids to form well characterised salts which contain one equivalent of acid. Of
these, the hydrochloride only possesses commercial im-

portance. *Aniline hydrochloride*, $C_6H_5NH_2 \cdot HCl$,* commercially known as "aniline salt," is obtained by mixing
pure aniline (100 pts.) with strong hydrochloric acid
(130–135 pts.) in stone-ware vessels. The salt is allowed
to crystallise out, and is freed from the mother liquors by
centrifugal force. Aniline hydrochloride forms colourless
crystals, which are easily soluble in water and alcohol and
which sublimate at 192°. With platinum chloride it forms
a double salt $(C_6H_5NH_2HCl)_2 \, PtCl_4$, which is sometimes
used in the quantitative determination of aniline.

Aniline salt is used in the dyeing and printing of aniline
black.

Acetanilide, $C_6H_5.NH.C_2H_3O$. If aniline is mixed with
glacial acetic acid, acetate of aniline is formed, but if
the mixture is boiled for some hours, a further change
takes place and acetanilide is formed, while at the same
time water is evolved—

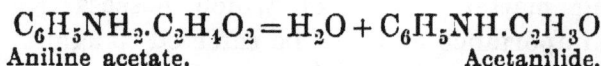

$$C_6H_5NH_2.C_2H_4O_2 = H_2O + C_6H_5NH.C_2H_3O$$

Aniline acetate.   Acetanilide.

Acetanilide forms colourless crystals, which melt at
115°. It boils without decomposition at 295°. It is used
in the manufacture of flavaniline.

Diamidobenzenes, $C_6H_4(NH_2)_2$.

The three isomeric diamidobenzenes which are possible
according to theory are formed by the complete reduction of the corresponding dinitrobenzenes.

* In the free amine bases the nitrogen is triatomic, but in their
salts it usually becomes pentatomic (as is ammonia salts). The
formula of aniline hydrochloride might therefore be represented
graphically thus—

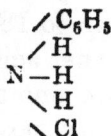

$$N \begin{cases} C_6H_5 \\ H \\ H \\ H \\ Cl \end{cases}$$

The most important of these is the *metaphenylene-diamine*, which is formed by the complete reduction of ordinary (meta) dinitrobenzene with iron and hydrochloric acid. It forms colourless crystals, which melt at 63° and dissolve easily in water. It boils at 287°. Meta phenylenediamine is used in the manufacture of chrysoidine and Bismarck brown.

Paraphenylenediamine is obtained by the reduction of paranitraniline (from paranitroacetanilide). It forms crystals, which melt at 147 and dissolve easily in water. When melted with sulphur it forms a thio compound which yields on oxidation Lauth's violet.

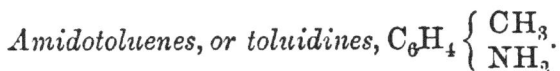

$$\textit{Amidotoluenes, or toluidines, } C_6H_4 \begin{cases} CH_3 \\ NH_2 \end{cases}.$$

The three isomeric toluidines may be obtained by the reduction of the corresponding pure nitrotoluenes. Commercial toluidine consists essentially of ortho and paratoluidines; metatoluidine occurs only in very small quantities and has little or no commercial interest. The separation of ortho and paratoluidine may be effected in several ways.

According to one method, the mixture of the bases is partially saturated with oxalic or sulphuric acid, and that which is left uncombined is driven over with steam. The principle of the method is that the paratoluidine combines with the acid first, and the orthotoluidine consequently distills over in the subsequent distillation.

According to another method, which has for its object the separation of orthotoluidine from paratoluidine and aniline, the mixture of the bases is neutralised with hydrochloric acid and mixed with a solution of phosphate of soda in excess. The contents of the vessel solidify, forming a crystalline paste, which dissolves on heating, the orthotoluidine floating to the surface as an oily layer, which is removed. On cooling the aqueous solution, the

H

aniline and paratoluidine separate out completely as phosphates, while a small quantity of phosphate of orthotoluidine remains in the mother liquor. The bases are now liberated by means of caustic soda, whereby the phosphate of soda is regenerated.

Orthotoluidine, $C_6H_4 \begin{Bmatrix} CH_3 \ (1) \\ NH_2 \ (2) \end{Bmatrix}$, forms a colourless liquid, which boils at 197–197·5°. Its specific gravity at 15° is 0·9978. It becomes brown when exposed to light and air, and dissolves in water in about the same proportion as aniline does. Orthotoluidine is a constituent of all commercial aniline oils except that which is used for black. In the pure state it is used in the manufacture of some azo dyes.

Metatoluidine, $C_6H_4 \begin{Bmatrix} CH_3 \ (1) \\ NH_2 \ (3) \end{Bmatrix}$, is a colourless oil, which boils at 197°.

Paratoluidine, $C_6H_4 \begin{Bmatrix} CH_3 \ (1) \\ NH_2 \ (4) \end{Bmatrix}$, forms colourless leaf-shaped crystals, which melt at 45°; it boils at 198°. It is only sparingly soluble in water, but dissolves easily in alcohol and ether. It is used in the manufacture of magenta.

Commercial products containing toluidine.

Aniline oil for red.—The aniline oil used in the manufacture of magenta consists of a mixture of the two toluidines (ortho and para) and aniline in about the proportion of one molecule of each. Its specific gravity may vary in oils destined to be used in the arsenic acid process from 1·001 to 1·010.

Aniline oil for safranine.—This product is the distillate (*échappé*) obtained in the manufacture of magenta. It consists essentially of aniline and orthotoluidine, and only contains a comparatively small quantity of paratoluidine.

Liquid toluidine consists of a mixture of ortho and paratoluidinc. It boils at 197–199°.

Tolylenediamines, $C_6H_3 (CH_3) (NH_2)_2.$

Of the six isomeric tolylenediamines possible according to theory, only five are known, and of these the most important is—

Alphatolylenediamine, $C_6H_3 \begin{cases} CH_3 \ (1) \\ NH_2 \ (2). \\ NH_2 \ (4) \end{cases}$ This compound is formed by the reduction of ordinary (1,2,4) dinitrotoluene. It forms colourless crystals, which melt at 99°, and dissolve easily in water. It is used, like metaphenylenediamine, in the manufacture of chrysoidine and Bismarck brown.

Amidoxylenes, or xylidines, $C_6H_3 (CH_3)_2 (NH_2).$

According to theory, six isomeric xylidines are possible, of which two are derived from orthoxylene, two from metaxylene, and one from paraxylene.

Commercial xylidine contains one amidoparaxylene, two amidoorthoxylenes, and two amidometaxylenes. It forms an oily liquid, which boils between 211° and 219°. Xylidine is used almost exclusively in the manufacture of the azo dyes.

Cumidines, $C_6H_2 (CH_3)_3 NH_2.$

According to theory, six isomeric amidotrimethylbenzenes are obtainable, of which only two have been properly investigated.

Commercial cumidine—

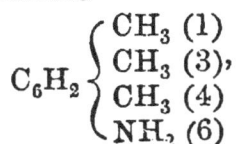

$$C_6H_2 \begin{cases} CH_3 \ (1) \\ CH_3 \ (3), \\ CH_3 \ (4) \\ NH_2 \ (6) \end{cases}$$

is obtained by heating xylidine hydrochloride with methyl alcohol in closed vessels to about 300°.

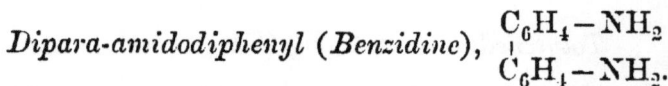

$$\text{Dipara-amidodiphenyl (Benzidine),} \quad \begin{array}{l} C_6H_4-NH_2 \\ | \\ C_6H_4-NH_2. \end{array}$$

This base, which was first obtained by Zinin in 1845, did not possess practical interest until the discovery of chrysamine and Congo red.

It is obtained by the reduction of azobenzene with stannous chloride in alcoholic solution, or by treating a solution of hydrazobenzene in warm hydrochloric acid with sulphurous acid. In the pure state it forms colourless shining crystals, which melt at 122°, and are almost insoluble in water. The sulphate is very sparingly soluble in water.

$$\text{Diamidoditolyl (tolidine),} \quad \begin{array}{l} C_6H_3(CH_3)NH_2 \\ | \\ C_6H_3(CH_3)NH_2 \end{array}, \text{ is obtained}$$

from orthohydrazotoluene, and forms colourless crystals, which melt at 128°. It is used, like the preceding compound, in the manufacture of azo dyes.

Amidonaphthalenes, or Naphthylamines, $C_{10}H_7NH_2$.

The two isomeric amidonaphthalenes possible according to theory are both known, and are of commercial importance.

Alphanaphthylamine, $C_{10}H_7NH_2$, is obtained by the reduction of alphanitronaphthalene with iron and hydrochloric acid. It forms colourless crystals, which possess a disagreeable smell, and which melt at 50°; the boiling point is 300°. With acids it combines to form well characterised salts, the aqueous solutions of which yield with mild oxidising agents a blue precipitate of naphthameïn. Alphanaphthylamine is used in the manufacture of azo dyes, naphthol yellow, and Magdala red.

Betanaphthylamine, $C_{10}H_2NH_2$. This amine is obtained on the large scale by heating the corresponding phenol or hydroxy compound with ammonia to a high temperature in closed vessels. The phenol used in this c ise is betanaphthol, and the reaction is expressed by the equation :—

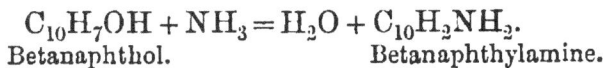

$$C_{10}H_7OH + NH_3 = H_2O + C_{10}H_7NH_2.$$

Betanaphthol.   Betanaphthylamine.

On the large scale, the following directions may be adhered to :—10 pts. betanaphthol are intimately mixed with 4 pts. caustic soda, and 4 pts. sal ammoniac, and heated for 60-70 hours in closed vessels to 150-160°.

Betanaphthylamine forms colourless crystals, which melt at 112°, and are devoid of smell. It is used chiefly in the manufacture of certain red azo dyes.

Secondary amines.

As has already been pointed out, the secondary amines may be regarded as derivatives of ammonia, NH_3, in which two hydrogen atoms have been replaced by organic radicals. They are formed in many ways, but the two chief reactions by means of which they are obtained on the large scale are the following :—

1. By heating equal molecules of a primary amine and its hydrochloride—

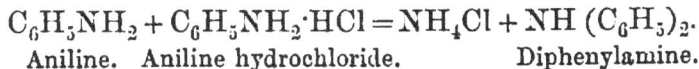

$$C_6H_5NH_2 + C_6H_5NH_2 \cdot HCl = NH_4Cl + NH(C_6H_5)_2.$$

Aniline.   Aniline hydrochloride.   Diphenylamine.

2. By heating primary aromatic amines with the chlorides of the alcohol radicals (methylchloride, ethylchloride, etc.)—

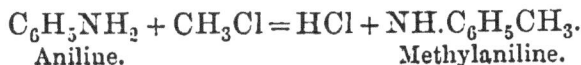

$$C_6H_5NH_2 + CH_3Cl = HCl + NH.C_6H_5CH_3.$$

Aniline.   Methylaniline.

On the large scale a modification of this reaction is employed, and in place of using the chloride of the alcohol radical the hydrochloride of the amine is heated with the alcohol in presence of zinc chloride—

$$C_6H_5NH_2.HCl + CH_3OH = NH.C_6H_5.CH_3 + HCl + H_2O.$$
Aniline hydrochloride.   Methylaniline.

The zinc chloride employed simply plays the part of a dehydrating agent.

Monomethylaniline, $C_6H_5NH.CH_3.$ or $N - CH_3 \Big\langle {}^{C_6H_5}_{H}$, is obtained on the large scale by heating together equal molecules of aniline hydrochloride and methyl alcohol to 200°. It forms a colourless liquid, which boils at 192°, and has a specific gravity at 15° of 0·976.

Monoethylaniline, $C_6H_5NH.C_2H_5$, is obtained in an analogous manner to that used for the preceding compound. It boils at 204°, and has a specific gravity at 18° of 0·954.

Diphenylamine, $NH(C_6H_5)_2$. On the large scale this compound is obtained by heating equal molecules of aniline and aniline hydrochloride to 220–230° in closed vessels. It forms colourless crystals, which melt at 54°, and which possess an agreeable smell, somewhat similar to that of violets. Diphenylamine is a weak base, the salts being decomposed by water. It is used in the manufacture of *diphenylamine blue, diphenylamine orange, and aurantia.*

Tertiary amines.

If all three hydrogen atoms in ammonia are replaced by organic radicals, tertiary amines are formed. This may be effected in the aromatic series by heating the primary amines with the chlorides of methyl, ethyl, propyl, etc.—

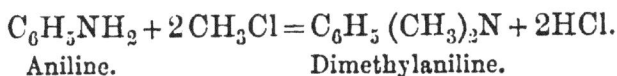

$$C_6H_5NH_2 + 2CH_3Cl = C_6H_5(CH_3)_2N + 2HCl.$$
Aniline.   Dimethylaniline.

According to another method, the hydrochlorides of the amines are heated with the fatty alcohols and zinc chloride—

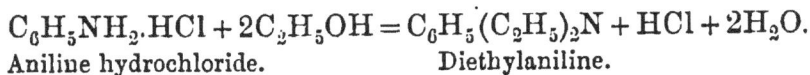

$$C_6H_5NH_2.HCl + 2C_2H_5OH = C_6H_5(C_2H_5)_2N + HCl + 2H_2O.$$
Aniline hydrochloride.   Diethylaniline.

The most important of the tertiary amines are:—

Dimethylaniline, $C_6H_5N(CH_3)_2$. For the preparation on the large scale, aniline and caustic soda are mixed in a closed vessel provided with a stirrer and pressure gauge. The contents of the vessel are heated to 100°, and the calculated quantity of methyl chloride is allowed to flow in, care being taken not to allow the pressure to exceed six atmospheres.

According to another method also employed on the large scale, 100 pts. aniline hydrochloride are treated with 50–80 pts. methyl alcohol, in closed enamelled vessels, to 180–200°. The dimethylaniline obtained according to this method contains more bye-products than that obtained by the preceding method.

Dimethylaniline is a colourless liquid, which solidifies at 0·5° and boils at 192. Its specific gravity at 15° is 0·96. It may be readily recognised by its property of forming methyl-violet, when heated with mild oxidising agents, such as copper chloride. Dimethylaniline is a very important commercial article, since it forms the chief raw material for the manufacture of a large number of coal-tar colours, including *methyl-violet*, *crystal-violet*, *malachite green*, *auramine*, *Victoria blue*, etc.

Nitrosodimethylaniline, $C_6H_4 \begin{cases} N(CH_3)_2 & (1) \\ NO & (4) \end{cases}$, is formed by the action of nitrous acid on dimethylaniline. For its preparation 10 pts. dimethylaniline are dissolved in 30 pts. hydrochloric acid and 200 pts. water. To the cooled solution a cold solution of 5·7 pts. sodium nitrite in 200 pts. water is added.

The free base forms green leaf-shaped crystals, which melt at 92°, and dissolve in water with a yellow colour. The salts are yellow. By reducing agents, nitrosodimethylaniline is converted into para-amidodimethylaniline, which on oxidation in presence of sulphuretted hydrogen yields *methylene blue*. When acted upon by phenols it yields a series of basic colours, which range from blue to violet, and are known, according to the discoverer, as *Meldola's colours*. Another colouring matter, known as *gallocyanin*, is obtained by heating together in solution nitrosodimethylaniline and tannic acid.

Diethylaniline, $C_6H_5N(C_2H_5)_2$, is formed by the action of ethylchloride on aniline, or by heating aniline hydrochloride with ethyl alcohol under pressure.

It forms a colourless liquid, which boils at 213·5° and has a sp. gravity of 0·936 at 18°. Its use is analogous to that of dimethylaniline.

Methyldiphenylamine, $(C_6H_5)_2N.CH_3$, is formed by heating diphenylamine hydrochloride with methyl alcohol under pressure to 250–300°. It forms a colourless oil, which boils at 282°, and which yields, when treated with oxidising agents, colours ranging from blue to violet. Heated with oxalic acid it forms *methyldiphenylamine blue*.

Pyridine bases.

These basic substances are found in coal-tar and in large quantities in the tar obtained by the distillation of animal substances (animal oil). Their simplest representative is pyridine itself, a body which possesses the empirical formula C_5H_5N. According to Körner (and this view has since been generally adopted), pyridine may be regarded as a benzene, in which one CH group has been replaced by an N atom, and the constitutional formula would therefore be :—

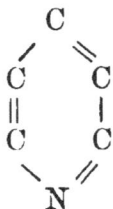

$$
\begin{array}{ccc}
& C & \\
\diagup & & \diagdown\!\!\!\diagdown \\
C & & C \\
\| & & | \\
C & & C \\
\diagdown & & \diagup\!\!\!\diagup \\
& N &
\end{array}
$$

Pyridine forms a colourless oil, which boils at 116°, and possesses a penetrating smell. Its homologue, methylpyridine, or *picoline*, $C_5H_4N.CH_3$, forms, according to theory, three isomers, two of which only are known.

These substances possess very little interest in the coal-tar colour industry.

Quinoline, C_9H_7N.

Quinoline bears the same relationship to naphthalene as pyridine to benzene. Its constitutional formula is represented thus :—

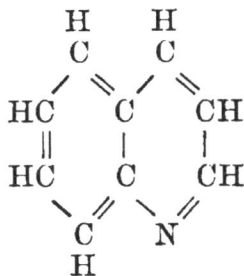

$$
\begin{array}{ccccc}
& H & & H & \\
& C & & C & \\
\diagup & & \diagdown\!\diagup & & \diagdown \\
HC & & C & & CH \\
\| & & | & & | \\
HC & & C & & CH \\
\diagdown & \diagup\!\diagup & & \diagdown\!\diagup & \\
& C & & N & \\
& H & &
\end{array}
$$

It is contained in coal-tar, and may be isolated from that portion of the *heavy oil* which is soluble in sulphuric acid. Quinoline is obtained, according to Skraup's synthesis, by the action of aniline and sulphuric acid on glycerin in presence of nitrobenzine—

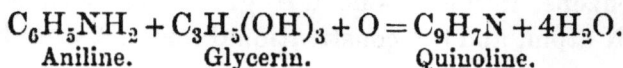

$$C_6H_5NH_2 + C_3H_5(OH)_3 + O = C_9H_7N + 4H_2O.$$
Aniline.   Glycerin.   Quinoline.

It forms a colourless oily liquid, which boils at 238° and has a sp. gravity of 1·094 at 20°. With the acids it forms usually well characterised salts.

Quinoline is used in the manufacture of *quinoline blue* (cyanin), *quinoline red* and *quinoline yellow*(quinophthalon).

Quinaldine, or methylquinoline, $C_9H_6N.CH_3$, occurs along with quinoline in coal-tar. The methyl group is in the ortho position to the N atom. It can also be obtained synthetically.

Acridine, $C_{13}H_9N$, bears the same relationship to anthracene as quinoline to naphthalene. It is almost invariably a constituent of crude anthracene.

Diazo compounds.

These peculiar substances are only known in the aromatic series. They all contain the group $-N:N-$ which is combined on the one side with a carbon atom of the benzene (naphthalene, etc.) ring, on the other with an acid radical or with a basic radical.

The diazo compounds are formed by the action of nitrous acid on the salts of the amido compounds in the cold. Formerly this was effected by passing nitrous fumes (from nitric acid and arsenious oxide) into the solution of the sulphate or chloride of the amine, but at present the calculated quantity of nitrite of soda and sulphuric acid are added. The formation of the nitrous acid is shown by the equation :—

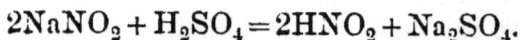

$$2NaNO_2 + H_2SO_4 = 2HNO_2 + Na_2SO_4.$$

This latter method is much more convenient than the old one, since, in the first place, no special apparatus is required; and secondly, the amount of nitrous acid required can be accurately gauged. On the large scale ice is usually added during the operation, or the liquid is cooled by other artificial means. The following equation expresses the formation of one of the simplest diazo compounds:—

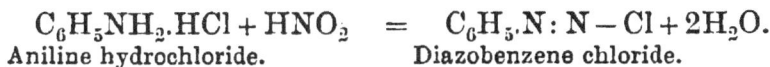

$$C_6H_5NH_2.HCl + HNO_2 = C_6H_5.N : N - Cl + 2H_2O.$$
Aniline hydrochloride.   Diazobenzene chloride.

The free diazo compounds are very unstable, and their salts, especially the nitrates, are in the dry state extremely explosive. They are never prepared on the large scale in the dry state, this being unnecessary. They are formed in solution, and are converted as a rule without delay into the corresponding oxyazo or amidoazo dyes.

When boiled in acid solution they are converted into the corresponding phenols with evolution of nitrogen—

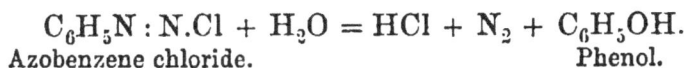

$$C_6H_5N : N.Cl + H_2O = HCl + N_2 + C_6H_5OH.$$
Azobenzene chloride.   Phenol.

When treated in solution with the salts of the amido compounds, the diazo compounds yield amidoazo compounds—

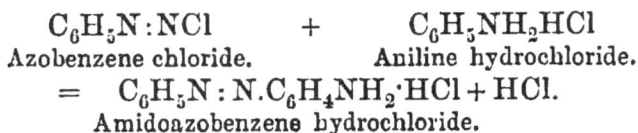

$$C_6H_5N:NCl + C_6H_5NH_2HCl$$
Azobenzene chloride.   Aniline hydrochloride.
$$= C_6H_5N : N.C_6H_4NH_2·HCl + HCl.$$
Amidoazobenzene hydrochloride.

With the phenols in alkaline solution oxyazo compounds are formed—

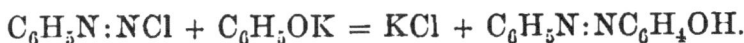

$$C_6H_5N:NCl + C_6H_5OK = KCl + C_6H_5N:NC_6H_4OH.$$

The two latter reactions are of great importance in the manufacture of the azo dyes.

Hydrazines.

The hydrazines are derived from the hypothetical radical NH_2-NH_2. The only compound of this class which appears to possess commercial importance is—

Phenylhydrazine, $NH_2-NH-C_6H_5$. It is obtained by reducing diazobenzene chloride in solution either with sodium sulphite and zinc dust, or with an acid solution of stannous chloride. Phenylhydrazine forms in the freshly prepared state a colourless oil possessing an aromatic odour. It melts at 23° and boils at 233°–234°. The hydrochloride $NH_2-NH-C_6H_5 . HCl$ crystallises in fine colourless leaflets.

The Azo compounds.

The azo compounds contain the group $-N:N-$ which is combined on both sides with carbon atoms of the benzene ring; *e.g.*, azobenzene, $C_6H_5-N:N-C_6H_5$. The simple azo compounds are obtained by the reduction of the primary amines in alkaline alcoholic solution—

$$2C_6H_5NO_2 + 4H_2 = C_6H_5N:N.C_6H_5 + 4H_2O.$$

By treating the diazo compounds in solution with salts of the amido compounds, or of the phenols, the *amidoazo compounds* and *oxyazo compounds* are obtained. This method, which has already been alluded to in speaking of the diazo compounds, is of great technical importance, since it forms the basis of the manufacture of the azo dyes.

The azo compounds are characterised by their great stability as compared with the diazo compounds. They are not changed by boiling in either acid or alkaline solution, and are not explosive. Some of them can even be distilled without undergoing decomposition. By mild reducing agents they are converted into the *hydrazo compounds*—

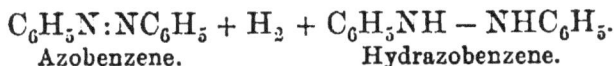

$$C_6H_5N:NC_6H_5 + H_2 + C_6H_5NH - NHC_6H_5.$$

Azobenzene.   Hydrazobenzene.

Strong reducing agents convert the $-N:N-$group into two amido groups. In this manner amidoazo compounds yield a monamine and a diamine—

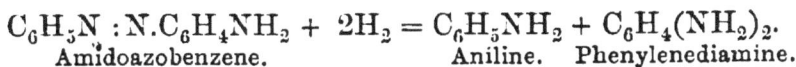

$$C_6H_5N :N.C_6H_4NH_2 + 2H_2 = C_6H_5NH_2 + C_6H_4(NH_2)_2.$$

Amidoazobenzene.   Aniline.   Phenylenediamine.

Oxyazo compounds yield under similar circumstances an amine and an amido phenol—

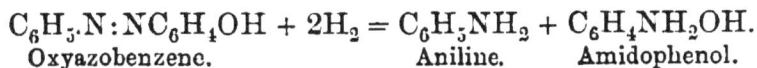

$$C_6H_5.N:NC_6H_4OH + 2H_2 = C_6H_5NH_2 + C_6H_4NH_2OH.$$

Oxyazobenzene.   Aniline.   Amidophenol.

This latter reaction is sometimes employed in finding the composition of azodyes.

The Sulphonic acids.

The aromatic sulphonic acids contain the group $-SO_3H$ in combination with one of the carbon atoms of the benzene ring. They are obtained almost exclusively by treating the aromatic hydrocarbons or their derivatives with strong sulphuric acid. Sometimes fuming sulphuric acid is used for the purpose. According to temperature and duration of the operation, either mono or disulphonic acids are formed—

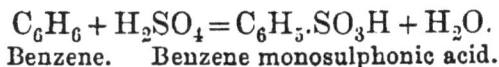

$$C_6H_6 + H_2SO_4 = C_6H_5.SO_3H + H_2O.$$

Benzene.   Benzene monosulphonic acid.

$$C_6H_6 + 2H_2SO_4 \quad = \quad C_6H_4(SO_3H)_2 + 2H_2O.$$

Benzenedisulphonic acid.

The trisulphonic acids are as a rule difficult to obtain, and are not of much commercial value.

The conversion of the aromatic compounds into sulphonic acids is technically known as the *sulphonation*, and the resulting substances are sometimes known as sulphonated products. Amines and phenols are as a rule more readily acted upon by sulphuric acid that the hydrocarbons. Amidosulphonic acids can sometimes be obtained even by heating the sulphate to about 180°. In the majority of cases the free sulphonic acids are not prepared, but in their place the sodium salts. The following is the general method followed for their isolation :—

After the action of the strong or fuming sulphuric acid has proceeded sufficiently far,* the product is poured into water, and the free sulphuric acid removed by saturating the liquid with barium carbonate, lead carbonate, chalk or milk of lime. On the large scale chalk or milk of lime are preferred for cheapness sake. The greater part of the free sulphuric acid is thus precipitated as gypsum, while the sulphonic acid combines with the lime to form a soluble lime salt. The liquid is now filtered hot, and the lime salt converted in the filtrate into the soda salt by the addition of the necessary amount of sodium carbonate. The following example (preparation of benzene monosulphonate of soda) will serve to illustrate the reactions which take place :—

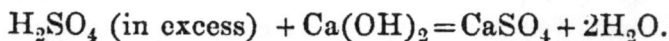

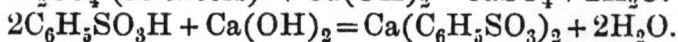

$$H_2SO_4 \text{ (in excess)} + Ca(OH)_2 = CaSO_4 + 2H_2O.$$
$$2C_6H_5SO_3H + Ca(OH)_2 = Ca(C_6H_5SO_3)_2 + 2H_2O.$$
Benzene monosulphonate of calcium.
$$Ca(C_6H_5SO_3)_2 + Na_2CO_3 = CaCO_3 + 2C_6H_5SO_3Na.$$

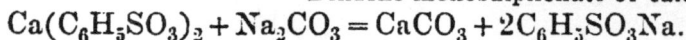

The liquid is now simply decanted or filtered from the precipitated calcium carbonate and evaporated to dryness.

---

* If sulphurous acid is evolved during the process of sulphonation, this is a sign that an oxidation and probably a more or less complete destruction of the molecule is taking place.

The pure sulphonic acids are as a rule deliquescent substances, difficult to crystallise, and not characterised by definite melting or boiling points. In order to characterise them, they are either converted into the corresponding chlorides or amides. The chlorides are best obtained by treating the free sulphonic acids with phosphorus pentachloride—

$$C_6H_5SO_3H + PCl_5 = C_6H_5SO_2Cl + POCl_3 + HCl.$$

The amides are obtained by the action of ammonia on the corresponding chlorides—

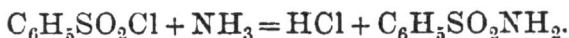

$$C_6H_5SO_2Cl + NH_3 = HCl + C_6H_5SO_2NH_2.$$

Benzene monosulphonic acid, $C_6H_5SO_3H$, forms colourless crystals, easily soluble in water and alcohol. By the action of caustic soda at a high temperature it is converted into phenol. With concentrated nitric acid it yields the isomeric nitrophenols. Treated in aqueous solution with nitrous acid it yields nitrosophenol.

Benzene disulphonic acids, $C_6H_4(SO_3H)_2$. All the three isomers are known. By the direct action of sulphuric acid (fuming) on benzene or benzene monosulphonic acid a mixture of the meta and paradisulphonic acids is obtained. This mixture is prepared on the large scale for the manufacture of resorcin either by passing the vapour of benzene into concentrated sulphuric acid heated to 240°, or by dissolving 1 pt. benzene in 4 pts. fuming sulphuric acid, and heating the solution of the monosulphonic acid obtained in this manner for two hours to 275°. When treated with caustic soda both these products are converted into metadioxybenzene (resorcin).

Toluene also yields sulphonic acids, which are, however, not of great importance.

Naphthalene monosulphonic acids, $C_{10}H_7(SO_3H)$. By heating naphthalene with an equal weight of concentrated

sulphuric acid both alpha and betanaphthalene monosulphonic acids are formed. By regulating the temperature it is possible to produce the one or the other of these
isomers in excess. Thus if the sulphonation is carried
out at 100° the product contains about 80 per cent. of
the alpha sulphonic acid and 20 per cent. of the beta
acid, whereas at a temperature of 160–170° about 75 per
cent. of the beta acid and 25 per cent. of the alpha acid
are obtained. The separation of the two products on the
large scale is based upon the different solubility of their
lime or lead salts, those of the beta acid being sparingly
soluble in water, while those of the alpha acid are easily
soluble. When melted with caustic soda the alpha acid
yields alphanaphthol, the beta acid betanaphthol.

Alphanaphthalene sulphonic acid, $C_{10}H_7(SO_3H)$ (α),
forms a crystalline deliquescent mass, which melts at
85–90°. Heated to 180° with dilute sulphuric acid it
is resolved into naphthalene and sulphuric acid. When
treated with concentrated sulphuric acid it is converted
into the isomeric beta acid.

Betanaphthalene sulphonic acid, $C_{10}H_7(SO_3H)$ (β),
forms leaf-shaped crystals, which do not deliquesce.

Nitrosulphonic acids. The three isomeric benzene nitro-
sulphonic acids are formed simultaneously when nitrobenzene is treated with fuming sulphuric acid, or when
benzene monosulphonic is treated with concentrated nitric
acid. The chief product of the reaction is in either case
metanitrobenzene sulphonic acid.

By the sulphonation of the nitronaphthalenes, or the
nitration of the naphthalene sulphonic acids, several
isomeric nitronaphthalene sulphonic acids have been
obtained.

Amidosulphonic acids. The amidosulphonic acids are
formed:—

1. By the direct action of strong sulphuric acid on the
amines.

2. In some cases by simply heating the sulphates of the amines to 180–230°.

3. By the reduction of the nitrosulphonic acids in acid solution.

Amidobenzene sulphonic acids, $C_6H_4 \begin{cases} NH_2 \\ SO_3H. \end{cases}$ All the three isomers are known. The meta and para compounds are of technical importance.

Meta-amidobenzene sulphonic acid, $C_6H_4 \begin{cases} NH_2 & (1) \\ SO_3H & (3) \end{cases}$, is obtained by the reduction of the corresponding nitrobenzene sulphonic acid with iron and dilute sulphuric acid. In the anhydrous state it forms long, fine, needle-shaped crystals, while from aqueous solution it crystallises with $1\frac{1}{2}$ mol. H_2O in clinorhombic crystals.

Para-amidobenzene sulphonic acid, $C_6H_4 \begin{cases} NH_2 & (1) \\ SO_3H & (4) \end{cases}$, commonly known as *sulphanilic acid*, is obtained by heating aniline (1 pt.) with strong sulphuric acid (3 pts.) to 180–190°. In the pure state it forms colourless crystals, which contain one molecule of water of crystallisation, which latter is given off on exposure to the air, the product falling to a white powder. It is sparingly soluble in water, and combines with both acids and bases to form salts.

Amidotoluene sulphonic acids, $C_6H_3 \begin{cases} CH_3 \\ NH_2 \\ SO_3H \end{cases}$. The *ortho-toluidine meta sulphonic acid,* $C_6H_3(CH_3.NH_2.SO_3H)$ (1.2.5), is obtained by the direct action of sulphuric acid on orthotoluidine, or by heating orthotoluidine sulphate to 200°. By the action of fuming sulphuric acid on paratoluidine two isomeric sulphonic acids are formed.

Xylidine sulphonic acids, $C_6H_2(CH_3)_2NH_2SO_3H$. Numerous isomers are possible, but the only one which

I

appears to possess technical importance is the product obtained by treating the ordinary xylidine of commerce (alpha-amidometaxylene) with concentrated sulphuric acid.

Benzidine sulphonic acids. By heating benzidine with fuming sulphuric acid, mono, di. tri, and tetra-sulphonic acids are formed. If the temperature is kept below 120°, the chief product of the reaction is benzidine-sulphon,

$$\left.\begin{array}{l} C_6H_3NH_2 \\ C_6H_3NH_2 \end{array}\right\} SO_2,$$

which is converted at 150–160° into sulphonic acids of benzidine sulphon, at higher temperatures into benzidine sulphonic acids.

Amidonaphthalene sulphonic acids.

The sulphonic acids of the two amidonaphthalenes can be obtained either by the direct sulphonation of the bases, or by heating the corresponding naphthol sulphonic acids with ammonia under pressure. Sulphonic acids of alpha-naphthylamine can also be obtained by the reduction of the alphanitronaphthalene sulphonic acids. The number of possible isomers is very great.

Sulphonic acids of alphanaphthylamine. By heating alphanaphthylamine with from 3 to 5 times its weight of strong sulphuric acid, or by heating alphanaphthylamine sulphate to 200°, a monosulphonic acid—

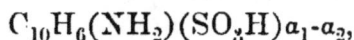

$$C_{10}H_6(NH_2)(SO_3H)a_1\text{-}a_2,$$

is obtained, which is known as *Naphthionic acid* (Piria's acid). It is not easily soluble in water, and crystallises with $1\frac{1}{2}$ mol. H_2O in small shining needles.

In addition to Piria's acid, the following *alpha naphthylamine monosulphonic acids* are also known* :—

* Nölting and Reverdin. *Bull de la Soc. ind. de Mulhouse*, 1887.

$C_{10}H_6(NH_2)(SO_3H)a_1-\beta_1$ by treating alphanaphthyl-amine hydrochloride with fuming sulphuric acid in the cold (Witt).

$C_{10}H_6(NH_2)(SO_3H)a_1 = \beta_1$. By reduction of the beta-nitronaphthalene sulphonic acid (Clève).

$C_{10}H_6(NH_2)(SO_3H)a_1 = a_3$. By the reduction of the alpha-nitronaphthalene sulphonic acid (Clève). This is the acid L of Schoellkopf & Co.

Two further isomers have been obtained by Clève, by the reduction of a nitronaphthalene sulphonic acid. Their constitution is not known.

By treating alphanaphthylamine or naphthionic acid with fuming sulphuric acid, several isomeric *alpha-naphthylamine disulphonic acids* are obtained, the constitution of which is not known.

The *sulphonic acids* of *betanaphthylamine* are of more importance in the manufacture of azo dyes than those of alphanaphthylamine. The following are known:—

$C_{10}H_6(NH_2)(SO_3H)$ β_1-a_1. (Alpha acid.) By treating betanaphthylamine with strong sulphuric acid (Bad. Anil. und Soda Fab.) (Dahl & Co.), or by the action of ammonia on the corresponding naphthol sulphonic acid.

$C_{10}H_6(NH_2)(SO_3H)$ $\beta_1 = a_1$. Acid γ III. of Dahl & Co.

$C_{10}H_6(NH_2)(SO_3H)$ $\beta_1 = \beta_1$. (Delta acid.) From betanaphthylamine and concentrated sulphuric acid at 170–180° (Bayer & Duisberg), or by the action of ammonia on Cassella's naphthol sulphonic acid.

$C_{10}H_6(NH_2)(SO_3H)\beta_1 = \beta_2$. Brönner's acid (beta acid) from Schaeffer's betanaphtholsulphonic acid and ammonia.

Quinoline monosulphonic acid, $C_9H_6N(SO_3H)$.

By the action of fuming sulphuric acid on quinoline, the ortho and meta sulphonic acids are formed. The para compound is obtained by heating sulphanilic acid (100 pts.) with glycerin (120 pts.), concentrated sulphuric acid

(150 pts.), and nitrobenzene (50–60 pts.). By heating the two first monosulphonic acids with fuming sulphuric acid to 200–240°, two isomeric disulphonic acids, $C_9H_5N(SO_3H)_2$, are formed.

The phenols.

The phenols, sometimes known as oxy compounds, are formed by the substitution of one or more of the hydrogen atoms of the benzene ring by the hydroxyl (OH) group.

These are formed :—

1. By the decomposition of organic substances at high temperatures, and are consequently contained in coal-tar.

2. By boiling the diazo compounds with dilute acids—

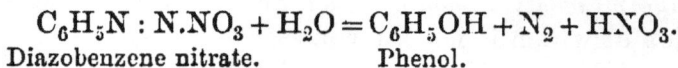

$$C_6H_5N : N.NO_3 + H_2O = C_6H_5OH + N_2 + HNO_3.$$

Diazobenzene nitrate.   Phenol.

Although this method is generally applicable, it is not used on the large scale, on account of the expense it incurs.

3. By melting the salts of the sulphonic acids or the halogen substitution derivatives with caustic soda or caustic potash—

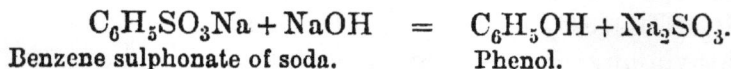

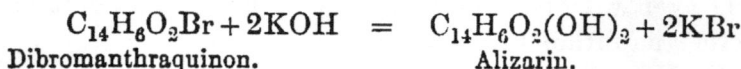

$$C_6H_5SO_3Na + NaOH = C_6H_5OH + Na_2SO_3.$$

Benzene sulphonate of soda.   Phenol.

$$C_{14}H_6O_2Br + 2KOH = C_{14}H_6O_2(OH)_2 + 2KBr$$

Dibromanthraquinon.   Alizarin.

The other modes of formation are of theoretical interest only.

The phenols are characterised by the following properties :—

The hydroxyl hydrogen is easily replaceable by metals or basic radicals—

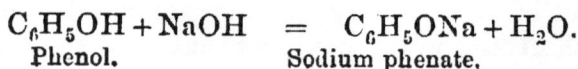

$$C_6H_5OH + NaOH = C_6H_5ONa + H_2O.$$

Phenol.   Sodium phenate.

This acid property, which is common to all the phenols, is increased by the introduction of halogen atoms or nitro groups into the benzene ring. Thus, whereas phenate of soda (C_6H_5ONa) can be decomposed by carbonic acid, the corresponding compound of trinitro-phenol (picrate of soda) is unacted upon.

The ethers of the phenols are most readily obtained by acting on the solution of the phenol in caustic alkali with the alkyl halogen derivatives—

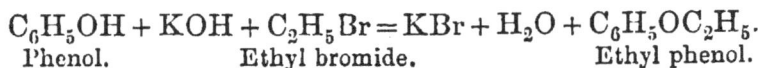

$$C_6H_5OH + KOH + C_2H_5Br = KBr + H_2O + C_6H_5OC_2H_5.$$
Phenol. Ethyl bromide. Ethyl phenol.

When distilled with zinc powder the phenols are converted into the corresponding hydrocarbons—

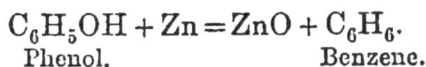

$$C_6H_5OH + Zn = ZnO + C_6H_6.$$
Phenol. Benzene.

The compounds of the phenols with soda or potash when acted upon at high temperatures with carbonic anhydride yield oxycarboxylic acids—

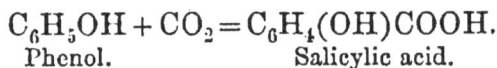

$$C_6H_5OH + CO_2 = C_6H_4(OH)COOH.$$
Phenol. Salicylic acid.

When brought into contact with diazo compounds in alkaline solution they form oxyazo compounds. This reaction is of great importance, as in this manner the majority of the azo dyes are obtained.

Nitrous acid readily acts upon the phenols, converting them into nitrosophenols.

When heated with ammonia they are more or less readily converted into the corresponding amines :—

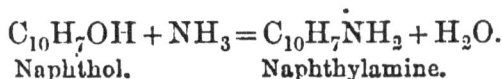

$$C_{10}H_7OH + NH_3 = C_{10}H_7NH_2 + H_2O.$$
Naphthol. Naphthylamine.

This reaction takes place much more readily in presence of dehydrating agents, such as zinc chloride or calcium chloride.

Phenol.—Carbolic acid or phenic acid, C_6H_5OH, was first discovered in 1834, by Runge, in coal-tar. It is obtained exclusively from coal-tar by the method described on p. 73.

In the pure state phenol forms long, colourless, prismatic crystals, which possess a powerful and characteristic smell. It melts at 42° and boils at 182°. At the ordinary temperature phenol dissolves in about fifteen times its weight of water; at 80–84° it will mix with water in all proportions. In alcohol and ether it is soluble in all proportions. When exposed to light and air it is soon coloured red. The aqueous solution gives with ammonia and chloride of lime a blue colouration, with ferric chloride a violet colouration, and with bromine water a yellowish precipitate of tribromphenol—

$$C_6H_2Br_3OH.$$

With concentrated nitric acid phenol yields, first mononitrophenols, then dinitrophenols, and ultimately trinitrophenol, or picric acid. With nitrous acid phenol yields nitrosophenol—$C_6H_4(OH)NO$. In concentrated sulphuric acid it dissolves with the formation of orthophenol sulphonic acid.

When heated with oxalic acid and concentrated sulphuric acid the rosolic acids are formed.

Phenol is used chiefly in the manufacture of picric acid, salicylic acid, rosolic acid, and as a disinfectant.

Phenate of soda, C_6H_5ONa, is obtained by dissolving phenol in caustic soda, and is used in the manufacture of salicylic acid.

Phenol methyl ether, or Anisol, $C_6H_5-OCH_3$, obtained by treating potassium phenate with methyl iodide or acid sulphate of methyl, forms a colourless liquid, which boils at 152°, and possesses an aromatic smell.

Dioxybenzenes.

All the three dioxybenzenes theoretically possible are known.

Pyrocatechin, $C_6H_4 \begin{cases} OH \\ OH \end{cases} \begin{vmatrix} 1 \\ 2 \end{vmatrix}$, is formed by the dry distillation of catechu; also by melting orthophenol sulphonic acid with caustic potash. It forms colourless crystals, dissolves easily in water, melts at 104° and boils at 245°. The aqueous solution gives an intense green colouration with ferric chloride.

The methyl ether, $C_6H_4 \begin{cases} OH \\ OCH_3 \end{cases}$, is contained in the tar obtained by the dry distillation of beechwood.

Resorcin, $C_6H_4 \begin{cases} OH \\ OH \end{cases} \begin{vmatrix} 1 \\ 3 \end{vmatrix}$, is obtained on the large scale by heating benzene disulphonate of soda with caustic soda. Resorcinate of soda is thus formed :—

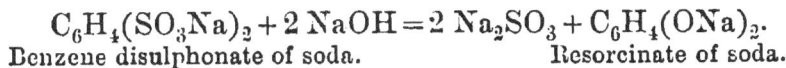

$$C_6H_4(SO_3Na)_2 + 2 NaOH = 2 Na_2SO_3 + C_6H_4(ONa)_2.$$
Benzene disulphonate of soda.   Resorcinate of soda.

The mass is then extracted with water, acidulated with sulphuric acid, in order to liberate the resorcin, and the aqueous solution evaporated down in shallow vessels. The crystals of sodium sulphate which separate out are removed from time to time by centrifugal action, and ultimately, when nearly all has crystallised out, a mother liquor remains, which contains nearly all the resorcin in aqueous solution. This is evaporated to dryness and distilled under reduced pressure.

Resorcin forms colourless crystals, which are easily soluble in water. The aqueous solution is neutral, possesses a sweet taste (poisonous), and soon turns red when exposed to the air. With ferric chloride it gives a dark violet; with chloride of lime, a violet colouration.

When melted with phthalic anhydride resorcin yields *fluorescein*, which dissolves in alkalies with an intense green fluorescence. This reaction is so delicate that it can be used as a test for resorcin, but it is not possible to distinguish it in this manner from its homologue cresorcin. When heated with citric acid in presence of concentrated sulphuric acid resorcin yields resocyanin, which dissolves in alkalies with a beautiful sky-blue fluorescence.

Hydroquinon, $C_6H_4{OH \atop OH} {|1 \atop |4}$, is obtained by oxidising aniline to quinon, reducing the solution with sulphurous acid and extracting with ether. It forms colourless crystals, which melt at 169° and which dissolve in seventeen parts of water.

Trioxybenzenes.

Three trioxybenzenes are possible according to theory, and all of these are known.

Pyrogallic acid, or pyrogallol, $C_6H_3{OH \atop OH}{OH}$ ${|1 \atop |3}{|2}$, was first obtained in 1786 by Scheele, by the dry distillation of gallic acid, but was considered by that chemist to be pure gallic acid.

It is prepared on the large scale by heating gallic acid with two to three times its weight of water, in closed vessels, to 200–210°, for about half an hour. The formation is represented by the equation:—

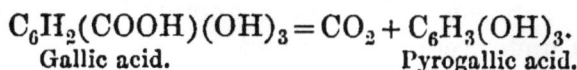

$$C_6H_2(COOH)(OH)_3 = CO_2 + C_6H_3(OH)_3.$$
Gallic acid.   Pyrogallic acid.

Pyrogallic acid forms white shining crystals, which melt at 131°. It boils at 210° with partial decomposition. In the dry state it remains unchanged on exposure to the air, but in alkaline solution it rapidly absorbs oxygen and assumes a dark or black colour. The aqueous solution is coloured red by ferric chloride. With ferrous sulphate containing small quantities of ferric salt it yields a deep indigo-blue colouration.

Pyrogallic acid is used in the manufacture of gallein, in photography, as a dye for the hair and for analytical purposes.

Phloroglucin, C_6H_3OH $\begin{matrix} OH & | & 1 \\ & | & 3, \\ OH & | & 5 \end{matrix}$ is best obtained by melting resorcin with caustic soda. It crystallises with two mol. H_2O in large colourless crystals.

Oxyhydroquinon, $C_6H_3(OH)_3$ (1, 2, 4), is formed by melting hydroquinon with caustic soda.

Oxytoluenes, or *cressols,* $C_6H_4(CH_3)(OH)$.

All the three isomers are contained in coal-tar, in which the ortho and para compounds exist in much larger quantities than the meta compound. They may be conveniently prepared in the pure state by transforming the corresponding amido compounds (toluidines) into diazo compounds and boiling these with dilute sulphuric acid.

Orthocressol melts at 32° and boils at 188°.

Metacressol is liquid at the ordinary temperature and boils at 201°.

Paracressol melts at 36° and boils at 199°.

The cressols show great similarity in their properties with phenol.

Dioxytoluenes, $C_6H_2CH_3(OH)_2$.

Orcin, $C_6H_3 \begin{cases} CH_3 & 1 \\ OH & 3 \\ OH & 5 \end{cases}$, may be obtained most readily by

extracting certain lichens (the Roccella tinctoria answers the purpose best) with milk of lime. The extract is boiled for some time and then treated with carbonic acid for the removal of the lime. The solution is then filtered and the filtrate evaporated to dryness on a water-bath. The residue is then crystallised—first from alcohol, and then from benzene.

There are, besides, numerous methods for the synthetical preparation of orcin. Thus it can be obtained by melting toluenemetadisulphonic acid or metadibromtoluene with caustic alkali.

Orcin crystallises with 1 mol. H_2O, in large colourless crystals, which melt at 58°. In the anhydrous state it melts at 106·5°–108° and boils at 290°. By the simultaneous action of air and ammonia orcin is converted into *orcein*, a colouring matter which forms the chief constituent of orchil and cudbear. Orcin dissolves easily in water.

Cresorcin $C_6H_3 \begin{cases} CH_3 & 1 \\ OH & 2 \\ OH & 4 \end{cases}$, is obtained by converting ortho-

amido paracressol into the diazo compound, and boiling this with dilute sulphuric acid. It forms colourless crystals, which dissolve easily in water and melt at 104–105°. Cresorcin shows a remarkable similarity to resorcin in its reactions.

Xylenols, $C_6H_3(CH_3)_2(OH)$. Six isomers are known.

Iso-butylphenol, $C_6H_4(CH_2 \cdot CH(CH_3)_2)(OH)$, is obtained by heating phenol with isobutyl alcohol and zinc chloride. It melts at 98°.

The higher oxy compounds in the benzene series are not of much technical importance.

The naphthols.

Both alpha and beta naphthol are contained in coal-tar, but in such small quantities that their isolation would not pay. They are both prepared on the large scale, the alpha compound being used chiefly in the manufacture of nitro-compounds, while the beta compound is used in the manufacture of azo dyes. They are prepared by melting the corresponding naphthalene sulphonic acids (q.v.) with caustic soda.

Alphanaphthol, $C_{10}H_7OH(a)$, forms lustrous monoclinic needles, which are sparingly soluble in water and which melt at 94°. It boils at 278–280°. A chip of pitch pine dipped into the aqueous solution of alphanaphthol and then into hydrochloric acid, soon turns green when exposed to sunlight, and ultimately brown. Ferric chloride yields in the aqueous solution a violet colouration, and on warming a violet precipitate of alphadinaphthol, $C_{20}H_{12}(OH)_2$.

Chlorate of potash and hydrochloric acid convert alpha-naphthol into dichlornaphthoquinon.

In the alkalies alpha-naphtol dissolves easily with formation of crystallisable naphtholates.

Alphanaphthol is rarely used in the manufacture of azo dyes. It is used chiefly in the manufacture of naphthol-yellow and as a disinfectant.

Betanaphthol, $C_{10}H_7OH$ (β), is obtained by heating a mixture of betanaphthalene sulphonic acid (1 pt.) with caustic soda (2 pts.), and just sufficient water to dissolve the caustic soda gradually to 300°. When the reaction is over, the mass is dissolved in water, acidulated with hydrochloric acid, and the liberated betanaphthol purified by distillation.

Betanaphthol forms in the pure state small lustrous crystals, which melt at 123°, and are only sparingly soluble in boiling water. It boils at 285–286°, with partial decomposition, can be sublimated, and distils over with steam. When treated with concentrated sulphuric acid ($1\frac{1}{2}$–2 parts) in the cold, betanaphthol yields betanaphthyl sulphate, $C_{10}H_7O.SO_3H$, analogous to acid ethyl sulphate. When heated, this compound undergoes a molecular change and is transformed into betanaphthol sulphonic acid. Heated with ammonia in closed vessels, beta-naphthol is easily transformed into betanaphthylamine.

The aqueous solution yields with chloride of lime solution a yellow colouration, with ferric chloride a slight green colouration. Nitrous acid yields with betanaphthol only one nitrosonaphthol, while with alphanaphthol two isomeric nitroso compounds are formed.

Naphtholsulphonic acids.

By treating the naphthols with strong sulphuric acid the first products of the reaction are naphthyl sulphates, which on further heating pass over into naphtholmono-sulphonic acids, of which numerous isomers exist. By the further action of sulphuric acid the naphthols are transformed into disulphonic and trisulphonic acids, of which a still greater number of isomers exists. In the preparation of these sulphonic acids on the large scale great care must be taken in regulating the temperature and time of sulphonating, as well as in the relative pro-portions of naphthol and sulphuric acid. Frequently several isomers are formed in the process of sulphonation, which may as a rule be separated by the different solu-bility of their salts. The sulphonic acids of the naph-thols, more especially those of betanaphthol, are of the greatest importance in the manufacture of the oxy-azo dyes, and their preparation and use for this purpose form the subject of numerous patents.

Sulphonic acids of alphanaphthol.

I. *Alphanaphtholmonosulphonic acids*, $C_{10}H_6OH.SO_3H$. The following are known :—

$C_{10}H_6OH.SO_3H$ a_1–β_1 is probably formed along with its isomer, a_1–a_2, by the action of sulphuric acid on alphanaphthol. It is obtained by the action of fuming sulphuric acid on a solution of alphanaphthol in glacial acetic acid. The ammoniacal solution shows a pure blue fluorescence.

$C_{10}H_6OH.SO_3H$ a_1–a_2 (Schaeffer's acid) is obtained by acting on alphanaphthol (1 pt.) with sulphuric acid (2 pts.) for two hours at 100°. The ammoniacal solution shows a violet fluorescence.

The acid having the constitution $a_1 = a_2$ forms an anhydride, $C_{10}H_6 \left\{ \begin{matrix} S\ O \\ O \end{matrix} \right\rangle$ at once. It is obtained by diazotising Schoellkopf's naphthylamine sulphonic acid and boiling the diazo compound in acid solution.

$C_{10}H_6OH.SO_3H$ $a_1 = \beta_1$. From betanaphthalene sulphonic acid and nitric acid (Clève).

$C_{10}H_6OH.SO_3H$ $a_1 = a_2$ from the corresponding naphthylamine sulphonic acid, treated with nitrous acid and boiled (Clève). Also by the reduction of the nitro compounds formed by alphanitronaphthalene and nitric acid or of the sulphonic acid formed from alphanaphthalene sulphonic acid and nitric acid.

$C_{10}H_6OH.SO_3H$ (Acid of Nevile and Winther). This acid, the constitution of which has not hitherto been determined, is obtained by diazotising naphthionic acid and boiling the diazo compound thus formed with dilute sulphuric acid. It is prepared in large quantities for subsequent use in the manufacture of azo-blue, brilliant scarlet, azo-orseillin, and azorubin S.

Alphanaphtholdisulphonic acids, $C_{10}H_5OH(SO_3H)_2$. Only one of the numerous possible isomers appears to be known. This is obtained by dissolving 1 pt. alphanaphthol in 3 pts. concentrated sulphuric acid, and heating the mixture for ten hours to 100–110°.

By dissolving alphanaphthol in fuming sulphuric acid, and treating the disulphonic acid thus formed with one mol. sodium nitrite in aqueous acid solution a nitroso-alphanaphthol disulphonic acid, $C_6H_4NO.OH (SO_3H)_2$, is formed, which dyes wool and silk a bright yellow in an acid bath (Seltzer).

Alphanaphtholtrisulphonic acid, $C_{10}H_4(OH)(SO_3H)(SO_3H)(SO_3H)$, $a_1 – \beta_1 – a_2 = a_2$, is obtained by the further sulphonation of Seltzer's acid.

Sulphonic acids of betanaphthol.

By the action of concentrated sulphuric acid on beta-naphthol at the ordinary temperature, naphthyl sulphate, $C_{10}H_7O.SO_3H$, is at first formed. On heating this gives rise to a mixture of betanaphtholmonosulphonic acids. The chief products of the reaction (at 100°) consist of two monosulphonic acids, which are known respectively as Schaeffer's and as Bayer's betanaphtholmonosulphonic acids. The relative quantities of the two acids formed depend upon the conditions observed during the process of sulphonation. Thus at 50–60° both acids are formed in about equal quantities, whereas at 100° Schaeffer's acid is formed in larger quantity.

By the further action of strong sulphuric acid on beta-naphthol, *disulphonic acids* are obtained. If one part beta-naphthol is heated with three parts strong sulphuric acid to 110°, there are formed (along with other substances) two betanaphtholdisulphonic acids, which are known respectively as betanaphtholdisulphonic acid R and betanaphtholdisulphonic acid G. The acid R yields with diazo compounds redder azo dyes than acid G.

By the action of fuming sulphuric acid on betanaphthol *trisulphonic acids* are formed,

Monosulphonic acids of betanaphthol.

$C_{10}H_6.OH.SO_3H$, β_1–a_1. This acid, which is generally known as Bayer's acid, is the chief product of the continued action of strong sulphuric acid on betanaphthol at the ordinary temperature. It has also been obtained by diazotising the corresponding naphthylaminesulphonic acid and boiling the resulting diazo compound in acid solution.

The free acid has not been obtained in the crystallised state. With diazo compounds it yields azo dyes, but in the case of diazoxylene it combines only in concentrated and not in dilute solution. With the bases it forms neutral and basic salts; thus with soda it forms a neutral sulphonate, $C_{10}H_6.OH.SO_3Na$, and a basic sulphate, $C_{10}H_6.ONa.SO_3Na$.

$C_{10}H_6OH.SO_3H$, $\beta_1 = \beta_2$. This acid, which is known as Schaeffer's acid, is formed along with the preceding by the action of strong sulphuric acid on betanaphthol at 50°–100°. For its preparation 1 pt. betanaphthol is dissolved in 2 pts. concentrated sulphuric acid, and the temperature maintained for some time at about 60°. The mass is then poured into water, neutralised with lime, filtered from the calcium sulphate formed, and the lime salts of the two isomeric sulphonic acids converted into the neutral sodium salts by the addition of sodium carbonate. The separation of the sodium salts may be effected either by treatment with boiling alcohol, in which the salt formed by Bayer's acid dissolves easily, while the salt of Schaeffer's acid remains behind, or by diluting the solution of the sulphonic acids in sulphuric acid with twice its weight of water, and then neutralising with sodium carbonate. In this manner four-fifths of Schaeffer's acid are separated out as a white crystalline precipitate. Another method of separation is based upon the fact that when a dilute ammoniacal solution of the two sulphonic acids is treated with diazoxylene, Schaeffer's acid at once combines to form colouring matter, while Bayer's acid remains unaltered.

Schaeffer's acid may also be obtained by diazotising the corresponding naphthylamine sulphonic acid, and decomposing the resulting diazo compound by boiling.

Schaeffer's acid crystallises in leaflets, which melt at 125°, are easily soluble in water and alcohol, and do not deliquesce.

$C_{10}H_6OH.SO_3H$, $\beta_1 = a_1$. The acid obtained by Dahl & Co. by the action of nitrous acid and boiling on γ naphthylamine sulphonic acid probably possesses this constitution (Nölting & Reverdin).

$C_{10}H_6.OH.SO_3H$, $\beta_1 = \beta_1$. From the alphanaphthalene disulphonic acid of Ebert & Merz, and caustic soda at 200–250°. The same compound is produced by the action of nitrous acid and boiling on the deltanaphthylamine sulphonic acid (Bayer & Duisberg).

Disulphonic acids of betanaphthol.

When betanaphthol is treated for several hours with 2-3 times its weight of concentrated sulphuric acid at 100–110°, the chief product of the reaction is found to consist of a mixture of two disulphonic acids, which are characterized respectively as betanaphtholdisulphonic acid R and betanaphtholdisulphonic acid G. This denomination has been chosen in order to express the fact that the former yields with diazo compounds red shades, while the latter yields yellow shades (German, *gelb*) of scarlet, etc. The two acids are also known as betanaphthol-A-disulphonic acid and betanaphthol-B-disulphonic acid respectively. The latter of these is also sometimes referred to in patent literature as the γ disulphonic acid of betanaphthol.

The separation of these two acids on the large scale forms the subject of numerous patents. According to Griess, they may be separated by the different solubility of their barium salts, that of the G acid being much more easily soluble in water than that of the R acid.

According to another method patented by Meister Lucius & Brüning, the separation may be effected by treating the mixture of the sodium salts (prepared in the usual way) with alcohol, in which that of the G acid dissolves, while that of the R acid remains undissolved.

Another method of separation patented by Beyer & Kegel is based upon the different behaviour of the sodium salts towards a saturated salt solution. By adding a saturated salt solution to the solution containing the two sodium salts, that of the R acid is separated or salted out, while that of the G acid remains in solution.

Betanaphtholdisulphonic acid, $R, C_{10}H_5OH(SO_3H)_2$, forms white silky crystals, which deliquesce rapidly in the air.

Betanaphtholdisulphonic acid, G, $C_{10}H_5OH(SO_3H)_2$, strongly resembles the former, but is more deliquescent.

Both acids are used extensively in the manufacture of azo dyes.

Betanaphtholtrisulphonic acid, $C_{10}H_4OH(SO_3H)_3$, is obtained by treating betanaphthol at 140-150° with 4-5 times its weight of a 20% solution of sulphuric anhydride in sulphuric acid.

Aromatic aldehydes.

The aromatic aldehydes contain, like those belonging to the fatty series, the group $- C \frac{= O}{- H}$.

They are formed :—

1. By the oxidation of the aromatic alcohols—

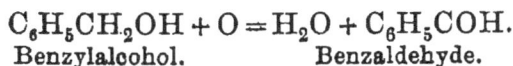

$$C_6H_5CH_2OH + O = H_2O + C_6H_5COH.$$
Benzylalcohol.   Benzaldehyde.

2. By the reduction of the aromatic acids—

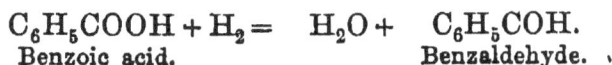

$$C_6H_5COOH + H_2 = H_2O + C_6H_5COH.$$
Benzoic acid.   Benzaldehyde.

3. By the action of boiling alkaline solutions on certain halogen disubstitution derivatives, in which the halogen is contained in the side chain—

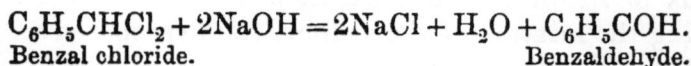

$$C_6H_5CHCl_2 + 2NaOH = 2NaCl + H_2O + C_6H_5COH.$$
Benzal chloride.   Benzaldehyde.

The aldehydes show the following characteristic reactions :—

By oxidising agents they are converted into the corresponding aromatic acids, while reducing agents produce the corresponding alcohols.

With alkaline bisulphites they yield double compounds. By ammonia and the amine bases they are acted upon with the formation of more or less complicated compounds.

Benzaldehyde, C_6H_5COH. This compound is by far the most important of the aromatic aldehydes from a commercial point of view. It was formerly obtained from bitter almonds, and is hence often known as oil of bitter almonds.

Benzaldehyde may be obtained on the large scale according to several methods. Thus it is formed, as stated above, by the action of boiling caustic soda on benzal chloride. It can also be obtained by boiling a mixture of benzal chloride and benzyl chloride ($C_6H_5CH_2Cl$) (obtained by passing chlorine into boiling toluene, until the specific gravity has reached 1·175) with water containing finely divided peroxide of manganese in suspension. According to another method, it is obtained by boiling benzyl chloride (1 part) with lead nitrate ($1\frac{1}{2}$ parts) and water (10 parts) in a vessel provided with a reflux condenser, and passing at the same time a current of carbonic acid through the liquid.

Benzaldehyde forms a colourless liquid, which refracts light powerfully and which boils at 179–180°. It has a specific gravity of 1·053 and is sparingly soluble in water. It possesses an agreeable smell, resembling that of bitter almonds. When exposed to the air it is rapidly oxidised and transformed into benzoic acid.

The commercial product usually contains benzoic acid, chlorinated benzaldehydes, and sometimes nitrobenzene, which latter substance must be regarded as an adulteration. The following are the usual tests applied :—When distilled, 90 volumes should pass over between 176 and 180°. 10 c.c. of the benzaldehyde should dissolve completely in 100 c.c. of a solution of sodium bisulphite (23° Tw.) at 40–50°. In order to test rapidly for nitrobenzene, two parts of the benzaldehyde are mixed with 1 pt. caustic potash; if nitrobenzene be present, the mixture will become green.

Benzaldehyde is used chiefly for the manufacture of the aniline greens; smaller quantities are used in making benzoic acid and cinnammic acid.

By the direct action of chlorine on benzaldehyde in presence of zinc chloride the meta and para substitution products $C_6H_4 \begin{cases} COH \\ Cl \end{cases}$ are formed.

Nitric acid produces ortho and meta nitrobenzaldehydes $C_6H_4 \begin{cases} COH \\ NO_2 \end{cases}$. The para compound is best obtained by oxidising the corresponding para nitrobenzyl chloride with lead nitrate. By reducing agents these three nitro compounds are converted into the corresponding amido compounds.

Cinnammic aldehyde, $C_6H_5 \cdot CH : CH \cdot COH$. This aldehyde is contained in oil of cinnamon and in oil of cassia. It may be prepared artificially by allowing benzaldehyde (10 pts.), ordinary aldehyde (15 pts.), water (900 pts.), and a ten per cent. solution of caustic soda (10 pts.), to stand with frequent agitation for 8–10 days at 30°. The liquid is then extracted with ether, and the residue left after distilling off the ether is subjected to dry distillation in a partial vacuum.

Cinnammic aldehyde forms a colourless oil, which is insoluble in and heavier than water. By the action of nitric acid it is converted into ortho-nitro cinnammic aldehyde (m.p. 127°) and para-nitro cinnammic aldehyde (m.p. 141–142°).

Ketons.

The ketons contain the group $\overset{|}{\underset{|}{C}} = O$, combined on either side with carbon atoms. They are formed :—

1. By the oxidation of the secondary alcohols—

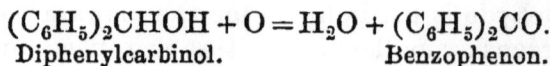

$$(C_6H_5)_2CHOH + O = H_2O + (C_6H_5)_2CO.$$

     Diphenylcarbinol.   Benzophenon.

2. By the oxidation of certain hydrocarbons—

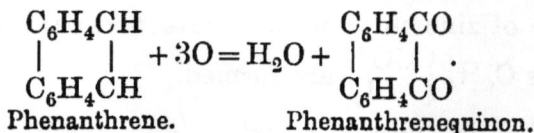

$$\begin{matrix} C_6H_4CH \\ | \quad\quad | \\ C_6H_4CH \end{matrix} + 3O = H_2O + \begin{matrix} C_6H_4CO \\ | \quad\quad | \\ C_6H_4CO \end{matrix}.$$

     Phenanthrene.   Phenanthrenequinon.

3. By distilling the lime salts of the aromatic acids, and besides, by several other methods, which are chiefly of scientific interest.

Benzophenon, $\left.\begin{matrix} C_6H_5 \\ C_6H_5 \end{matrix}\right\}$ CO, is best obtained by the dry distillation of benzoate of lime. It is insoluble in water, but easily soluble in alcohol and ether, and forms crystals, which melt at 27° or 49°, and boil at 300°. Benzophenon itself is not of much importance, but its amido derivatives, especially the tetramethyldiamidobenzophenon, form the raw materials for the manufacture of certain important colouring matters.

Tetramethyldiamidobenzophenon, $CO \begin{matrix} \diagup C_6H_4N(CH_3)_2 \\ \diagdown C_6H_4N(CH_3)_2 \end{matrix}$. This base is obtained on the large scale by the action of carbon oxychloride on dimethylaniline—

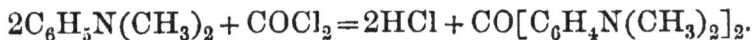

$$2C_6H_5N(CH_3)_2 + COCl_2 = 2HCl + CO[C_6H_4N(CH_3)_2]_2.$$

For its manufacture carbon oxychloride is passed into dimethylaniline until the increase in weight corresponds to the above equation. The crystalline paste thus obtained is heated for some time in closed vessels on a water-bath. When the reaction is over, the mass is dissolved in water, the unchanged dimethylaniline driven off by steam, and the residue is purified by dissolving in hydrochloric acid and precipitating with caustic soda.

Tetramethyldiamidobenzophenon forms colourless leaflets, which melt at 170° and are easily soluble in ether and hot alcohol. When heated with salts of ammonia or of the primary amines, in presence of zinc chloride, it yields yellow and orange colouring matters (Auramines).

Tetra-ethyldiamidobenzophenon resembles in its properties the preceding compound and is obtained in a similar manner from diethylaniline.

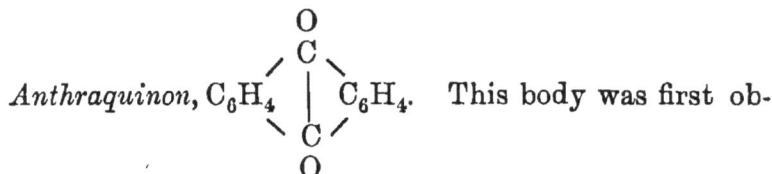

Anthraquinon, $C_6H_4 \begin{matrix} O \\ \diagup C \diagdown \\ | \quad | \\ \diagdown C \diagup \\ O \end{matrix} C_6H_4.$ This body was first ob-

tained by Laurent in 1840 by the oxidation of anthracene with nitric acid. The oxidation is at present effected by means of bichromate of potash and sulphuric acid. For its manufacture on the large scale the crude anthracene, containing 60–80% of pure substance, is first subdivided as finely as possible by sublimation and grinding. 100 kilos. anthracene are then added to a boiling solution of 100–150 kilos. bichromate of potash in 1,500 litres of water contained in a large lead-lined vessel. To the hot solution are then added, during 9-10 hours, 140–210 kilos. sulphuric acid at 168° Tw. previously diluted. The temperature generated by the reaction is sufficient to keep the liquid boiling during the operation. After all the acid has been added, the boiling is continued for a short period, when the whole is allowed to cool and the crude anthraquinon is separated by filtration or centrifugal force. The mother liquors which contain all the chromium, in the form of chrome alum, are not wasted, but are treated separately for the regeneration of bichromate. At this stage of the manufacture the anthraquinon appears as a reddish-yellow powder. For its further purification, it is heated in cast-iron vessels provided with stirrers with 2–3 pts. strong sulphuric acid to 80° and the temperature then gradually raised to 110° until all the anthraquinon is dissolved, and a sample of the liquid poured into water gives a pure white precipitate. On allowing the sulphuric acid solution to cool, part of the anthraquinon separates out in a crystalline form; the rest is precipitated by diluting the solution in leaden vessels with twenty times its volume of water. The precipitate, which contains about 90 per cent. anthraquinon, is collected in filter presses and presents now the appearance of a light-grey or yellowish-green crystalline powder. It may be still further purified by boiling with soda. The average yield from 100 pts. of 60 per cent. anthracene is from 50–55 pts. anthraquinon.

Anthraquinon, the constitution of which is represented by the formula :—

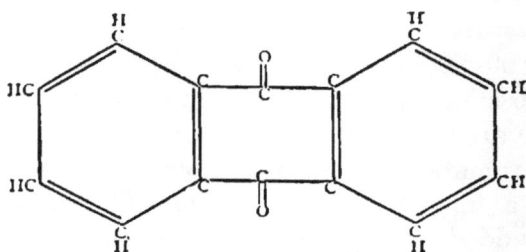

forms golden-yellow needles or prisms which melt at 277°. It is insoluble in water, and only sparingly soluble in alcohol and ether. It can be distilled without undergoing decomposition, and begins to sublimate in needles below its melting point. It dissolves in strong warm sulphuric acid, and the solution can be heated to 130° without any chemical action taking place, but at higher temperatures it is transformed into the mono- and di-sulphonic acids.

Anthraquinonmonosulphonic acid, $C_{14}H_7(SO_3H)O_2$, is obtained by the action of fuming sulphuric acid (containing about 50 per cent. anhydride) on anthracene at 160°.

The free acid forms fine yellow leaflets, and is easily soluble in water. When melted with caustic soda it yields alizarin.

Anthraquinondisulphonic acids, $C_{14}H_6(SO_3H)_2O_2$. Two disulphonic acids are known which are designated respectively as the alpha and beta compounds. By dissolving anthracene in fuming sulphuric acid of the same strength as that used in the preparation of the monosulphonic acid, and keeping the temperature for a considerably longer period at 160–170°, the chief product of the reaction will be the betasulphonic acid which crystallises with one molecule of water of crystallisation in fine golden yellow leaflets.

By raising the temperature for a short time to 180–185°, the alphadisulphonic acid is obtained as chief product. It also contains one molecule of water, and forms a golden yellow crystalline mass.

The alphasulphonic acid yields, when melted with caustic soda, flavopurpurin ; while the beta sulphonic acid treated in a similar manner gives rise to anthrapurpurin.

Aromatic Acids.

The aromatic acids contain, like other organic acids (carboxylic acids), the carboxyl group $-CO\cdot OH$. They may be mono, di, or tri-basic, etc., according to the number of such carboxyl groups contained in the molecule.

They are formed by the oxidation of the aromatic alcohols and aldehydes and of the aromatic hydrocarbons which contain one or more methyl, ethyl, etc., groups attached to the carbon atoms of the benzene ring. The following are the only bodies of this class which have any great importance in the coal-tar colour industry.

Benzoic acid, $C_6H_5 \cdot COOH$. This compound was discovered as far back as 1608, by Blaise de Vigenère, who obtained it by heating gum benzoë. Benzoic acid occurs, besides, naturally in several other resins and balsams, and as a constituent of many plants.

Small quantities of benzoic acid are still prepared from gum benzoë, but only for pharmaceutical purposes. According to v. Rad, artificial benzoic acid is best obtained by heating benzotrichloride with water under pressure. The conversion takes place according to the following equation :—

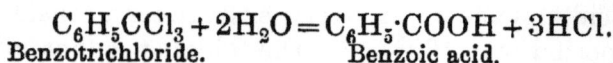

$$C_6H_5CCl_3 + 2H_2O = C_6H_5 \cdot COOH + 3HCl.$$
Benzotrichloride.   Benzoic acid.

Pure benzoic acid forms long white needle-shaped crystals which melt at 120° and boil at 250°. It sublimates easily below the boiling point, emitting a vapour which rapidly irritates the mucous membranes and causes coughing. It is sparingly soluble in water, but dissolves readily in alcohol and ether. With chlorine, bromine, nitric acid, and sulphuric acid it yields substitution derivatives of which those containing the substituting element or group in the meta-position to the carboxyl group are the chief products of the reaction.

Benzoic acid is employed in the manufacture of aniline blue and for pharmaceutical purposes.

Salicylic acid or *orthoxybenzoic acid*, $C_6H_4 \begin{cases} COOH(1), \\ OH \quad (2) \end{cases}$ occurs naturally in the blossom of the Spiræa ulmaria, and

as methyl ether in the oil obtained from Gautheria pro-
cumbeus (winter-green oil).

Salicylic acid is obtained artificially by heating phenate
of soda in a current of dry carbonic acid gradually from
100° to 200° during 6–8 hours. The grey mass of salicylate of soda thus obtained is dissolved in water, and
treated first with a small quantity of sulphuric acid, in
order to precipitate resinous and coloured impurities.
Excess of acid is then added in order to precipitate
all the salicylic acid, which is further purified by crystallisation. According to another method, salicylic
acid may be obtained by heating diphenyl carbonate
(obtained by the action of carbonoxychloride on phenol)
to 200° in an indifferent atmosphere.

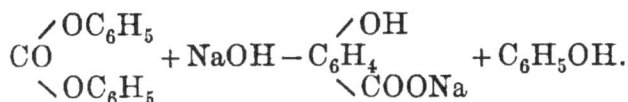

$$CO\begin{cases}OC_6H_5 \\ OC_6H_5\end{cases} + NaOH - C_6H_4\begin{cases}OH \\ COONa\end{cases} + C_6H_5OH.$$

Salicylic acid crystallises in long needle-shaped crystals,
which sublimate easily when heated gently, but when
heated rapidly decompose into phenol and carbonic acid.
It is only sparingly soluble in cold water, but dissolves
readily in boiling water, alcohol, and ether. The aqueous
solution gives with ferric chloride a violet colouration.
Like carbolic acid, free salicylic acid possesses the properties of a strong antiseptic. The solution in soda
possesses an intensely sweet taste, but no longer possesses
antiseptic properties. It is used in the manufacture of
chrysamine and other azo dyes, for surgical and medicinal
purposes, for the manufacture of artificial winter-green
oil, and in large quantities as an antiseptic in checking
fermentation, putrefaction, etc.

Gallic acid, $C_6H_2(OH)_3 \cdot COOH$, and gallotannic acid,
$C_{14}H_{10}O_9$, also belong to the aromatic acids. They are
not obtained from coal-tar, but are used as raw products
in the manufacture of one or two artificial colouring

matters.   Gallic acid occurs in many natural tannins, and
may also be obtained by allowing tannic acid to ferment
or by boiling it with dilute sulphuric acid. It dissolves in
100 pts. cold water, easily in alcohol and ether, and crystallises from water with one molecule of water of crystallisation.   It melts at 222°, and is resolved at higher
temperatures into pyrogallic acid and carbonic acid.
Gallic acid is not precipitated from its solutions by
gelatine.

Tannin or tannic acid forms a yellowish amorphous
powder, which is easily soluble in water, but insoluble in
ether. It is precipitated from its aqueous solution by
gelatine, and yields with salts of iron intense blue-black
precipitates (ink).

Phthalic acid, $C_6H_4 \begin{cases} COOH(1) \\ COOH(2) \end{cases}$.   This substance is
formed by the oxidation of many aromatic substances.
For its preparation, naphthalenetetrachloride, dinitronaphthol, or similar substances are oxidised by chromic
or nitric acid.

Phthalic acid is sparingly soluble in cold, easily in hot
water, and crystallises in short prisms or leaflets which
melt at 213°; at 130°, however, they are already converted
into phthalic anhydride.

Phthalic anhydride, $C_6H_4 \begin{cases} CO \\ CO \end{cases} O$.   This compound,
which is commercially known as phthalic acid, is obtained
from the preceding compound by sublimation, in long white
flexible needles. It melts at 128° and boils at 284°·5.
When boiled for a length of time with water it is reconverted into phthalic acid.   When heated with the phenols
(sometimes in presence of dehydrating agents) it yields
an important and interesting class of bodies known as
the phthaleïns, some of which are colouring matters in
themselves, whereas others (*e.g.*, fluoresceïn) form the raw
products for the manufacture of certain colours.

PART III.

THE COAL-TAR COLOURS.

THE coal-tar colours are usually divided into a number of groups, each of which deals with those products which are obtained from the same constituents of tar. Thus we have :—

1. Colouring matters from benzene and toluene.
2. Phenol colouring matters.
3. Naphthalene colouring matters.
4. Anthracene colouring matters.

The materials used in the preparation of many colouring matters are, however, sometimes derived from two, sometimes from three constituents of coal-tar. Thus, the eosins are obtained from phthalic acid and resorcin. The former is again obtained from naphthalene, while the latter is obtained from benzene. Eosin can therefore either be classed among those colouring matters which are obtained from benzene, or among the naphthalene colouring matters. Resorcin itself is a phenol, so that eosin might also be regarded as a phenol dye.

In order to overcome these difficulties, and especially in order to describe all those colouring matters which possess similar properties (although they may have originated from different sources) under one heading, the following classification of the most important colouring matters will be adhered to in this work :—

I. *Aniline dyes.*
  (*a*) Rosaniline group.
  (*b*) Indulines and Safranines.
  (*c*) Oxazines.
  (*d*) Aniline-black.
  (*e*) Colouring matters containing sulphur (thionines).

II. *Phenol dyes.*
  (*a*) Nitro bodies.
  (*b*) Colouring matters which are formed by the action of nitrous acid on the phenoles.
  (*c*) Rosolic acid.
  (*d*) Phthaleins and Indophenols.

III. *Azo dyes.*
  (*a*) Amidoazo dyes.
  (*b*) Amidoazo sulphonic acids.
  (*c*) Oxyazo dyes.

IV. Artificial indigo.

V. Anthracene dyes.

I. ANILINE DYES.

The denomination "aniline dyes" is not applicable to all the colouring matters described in this section. The term "amine dyes" would perhaps be more suitable, as it would include dye-stuffs from naphthylamine, as well as from aniline and toluidine; "aniline dyes" of known constitution may indeed be regarded as complicated amines. In the following, all dyes are described as "aniline dyes" which contain nitrogenous bases, or their derivatives,—for example, their sulphonic acids,—with the only exception of those which can with certainty be called "azo dyes."

(*a*) THE ROSANILINE GROUP.

The most important colouring matters of this group may be regarded as derivatives of two hydrocarbons:

triphenylmethane, $C_{19}H_{16}$, and tolyldiphenylmethane, $C_{20}H_{18}$. Their constitution is expressed by the following formulæ:—

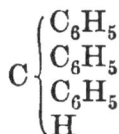

$$C\begin{cases} C_6H_5 \\ C_6H_5 \\ C_6H_5 \\ H \end{cases}$$

Triphenylmethane.

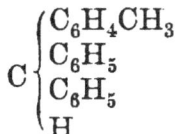

$$C\begin{cases} C_6H_4CH_3 \\ C_6H_5 \\ C_6H_5 \\ H \end{cases}$$

Tolyldiphenylmethane.

The carbon atom which joins the three benzene rings to each other is distinguished as "methane carbon."

If one hydrogen atom in two or three phenyl groups is replaced by NH_2, and further, the hydrogen of the NH_2 group is replaced by methyl, benzyl, phenyl, etc., we obtain a series of nitrogenous compounds; e.g.:—

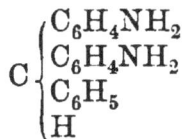

$$C\begin{cases} C_6H_4NH_2 \\ C_6H_4NH_2 \\ C_6H_5 \\ H \end{cases}$$

Diamidotriphenylmethane.

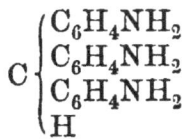

$$C\begin{cases} C_6H_4NH_2 \\ C_6H_4NH_2 \\ C_6H_4NH_2 \\ H \end{cases}$$

Triamidotriphenylmethane.
(Para-leucaniline.)

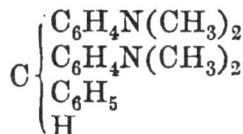

$$C\begin{cases} C_6H_4N(CH_3)_2 \\ C_6H_4N(CH_3)_2 \\ C_6H_5 \\ H \end{cases}$$

Tetramethyldiamidotriphenylmethane.
(Leuco-base of malachite-green.)

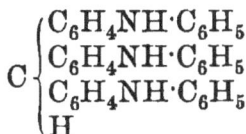

$$C\begin{cases} C_6H_4NH \cdot C_6H_5 \\ C_6H_4NH \cdot C_6H_5 \\ C_6H_4NH \cdot C_6H_5 \\ H \end{cases}$$

Triphenylleucaniline.

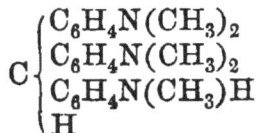

$$C\begin{cases} C_6H_4N(CH_3)_2 \\ C_6H_4N(CH_3)_2 \\ C_6H_4N(CH_3)H \\ H \end{cases}$$

Pentamethylleucaniline.

These compounds, called "leuco-bases," are colourless, and yield colourless salts with acids. By oxidation they are transformed, more or less readily, into the colour-bases, which differ from the "leuco-bases" by containing one atom of oxygen.

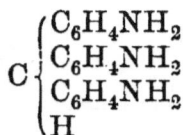

$$C \begin{cases} C_6H_4NH_2 \\ C_6H_4NH_2 \\ C_6H_4NH_2 \\ H \end{cases} \qquad\qquad C \begin{cases} C_6H_4NH_2 \\ C_6H_4NH_2 \\ C_6H_4NH_2 \\ OH \end{cases}$$

Para-leucaniline.   Para-rosaniline.

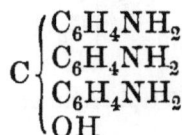

The colour-bases are generally colourless, or nearly so. They unite with acids, with elimination of water to form coloured salts, the real dye-stuffs. The manner in which this splitting off of water takes place has not been satisfactorily explained for all colouring matters. In the case of para-rosaniline it takes place according to the equation :—

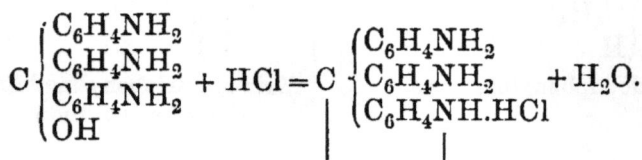

$$C \begin{cases} C_6H_4NH_2 \\ C_6H_4NH_2 \\ C_6H_4NH_2 \\ OH \end{cases} + HCl = C \begin{cases} C_6H_4NH_2 \\ C_6H_4NH_2 \\ C_6H_4NH.HCl \end{cases} + H_2O.$$

Beside these mono-acid, coloured salts, the bases yield a series of compounds containing two or more equivalents of acid, which are generally yellow or brown. Thus, hydrochloride of rosaniline, or magenta, $C_{20}H_{19}N_3HCl$, will unite with two further molecules of hydrochloric acid to form the triacid salt, $C_{20}H_{19}N_3.3HCl$.

This explains the behaviour of the colouring matters of this group in aqueous solution, or on the fibre, towards concentrated acids. Decolourisation takes place, but on adding water, the acid salt is decomposed, and the original colour is restored.

There exists also, in dilute solutions, a certain attraction between the normal salts (with one molecule of hydrochloric acid) and the free acid, and the consequence is that, in dyeing, the colouring matters of this group are either not absorbed at all, or only very imperfectly so, from a bath acidified with a mineral acid. For this reason these dyes must be dyed in a neutral or only very slightly acid bath.

The colouring matters of this group unite with the fibres as salts, and not as free bases. This is conclusively proved, in the first place, by the fact that when wool, for instance, is heated in a *colourless* solution of magenta-base, the fibre is soon dyed an intense red, which can only result from the combination of the magenta-base with some acid constituent of the fibre to form a coloured insoluble salt or lake. Furthermore, it has been shown than when wool or silk is dyed in neutral solutions of the basic dyes (hydrochlorides), such as magenta, crystal violet, chrysoidine, etc., *all* the hydrochloric acid remains in the solution, and the only conclusion which can be drawn is that a double decomposition has taken place between a constituent or constituents of the fibre and the hydrochloride of the colour-base. (See *Journ. Soc. Dyers & Col.*, 1888, p. 72.) They give with some acids precipitates insoluble in water (colour-lakes). The tannic acid lakes are soluble in alcohol. They also possess to some extent the property of uniting chemically with basic oxides, such as stannic oxide and alumina.

Thus, magenta may be fixed on cotton, if it is printed with acetate or arseniate of alumina and then steamed, but the colour is not fast. The weak affinity which fibres prepared with mineral mordants possess for some aniline dyes may, however, be ascribed in most cases to physical influences.

Malachite-green.

(Solid Green, Victoria Green, New Green, Benzal Green, Benzoyl Green, Fast Green.)

The colouring matter prepared by heating benzotrichloride and dimethylaniline, is known as malachite-green. Technically, the following method is more advantageous.

Benzaldehyde is heated with dimethylaniline and zinc chloride in alcoholic solution. The leuco-base of the green is thus formed :—

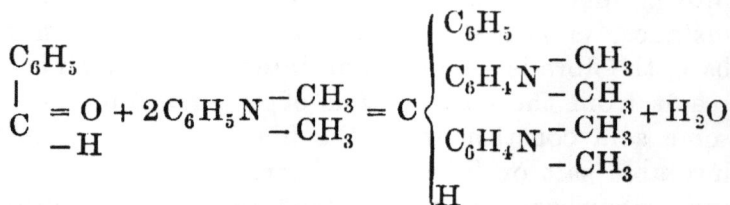

$$C_6H_5\underset{-H}{\overset{|}{C}}=O + 2\,C_6H_5N\overset{-CH_3}{\underset{-CH_3}{}} = C\begin{cases} C_6H_5 \\ C_6H_4N\overset{-CH_3}{\underset{-CH_3}{}} \\ C_6H_4N\overset{-CH_3}{\underset{-CH_3}{}} \\ H \end{cases} + H_2O$$

Benzaldehyde.   Dimethylaniline.   Tetramethyldiamido-
                                   triphenylmethane.

The leuco-base is precipitated with water, freed from unchanged dimethylaniline, dissolved in hydrochloric acid, and oxidised with lead peroxide. The free colour-base has the formula—

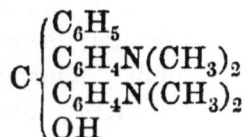

$$C\begin{cases} C_6H_5 \\ C_6H_4N(CH_3)_2 \\ C_6H_4N(CH_3)_2 \\ OH \end{cases}$$

The manner in which it unites with acids, with elimination of water, is not yet explained. Probably the hydrochloride has the formula :—

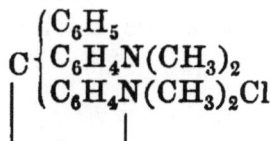

$$C\begin{cases} C_6H_5 \\ C_6H_4N(CH_3)_2 \\ C_6H_4N(CH_3)_2Cl \end{cases}$$

The hydrochloride is not well defined, and more difficult to separate than the oxalate and zinc double salt, both of which are commercial articles.

Properties.—The oxalate has the composition—

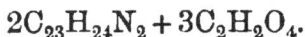

$$2C_{23}H_{24}N_2 + 3C_2H_2O_4.$$

It forms tablets possessing a green metallic lustre, which dissolve easily in water, alcohol, and amyl alcohol.

The zinc double salt, $3(C_{23}H_{24}N_2HCl) + 2ZnCl_2 + 2H_2O$, also crystallises well, with a yellow-green cantharides lustre. An inferior quality consists of the ferric chloride compound with the hydrochloride of the base.

The picrate is insoluble in water, and comes into commerce under the name of *Malachite green soluble in spirit.*

By adding caustic potash solution to a solution of one of the above salts, the free base separates out. It crystallises from petroleum spirit in colourless needles, which melt at 126° to 130°.

The solutions of the colouring matter are bluish-green; concentrated acids turn them orange-yellow, but the original colour is restored on dilution. *Stannous chloride* produces a green precipitate, chloride of lime decolourises the solution.

Application.—*Silk* is dyed in a pure soap bath, and brightened with acetic acid. *Wool* is dyed in a very weak acid bath. If too much acid is present, the colouring matter is not taken up completely. For dyeing and printing cotton, tannin is used, along with alumina mordants, or tartar emetic.

This green will not stand either soaping or milling, nor is it fast to light.

It may be detected on the fibre by the reaction with soap, or the orange colour with hydrochloric acid, which is restored by water. Acetic acid removes the colouring matter with a bluish-green tint, ammonia and alkalies

L

completely decolourise. By heating, the colour does not change to violet. (Distinction from methyl-green.)

Other greens belonging to this class are Ethyl-green, Solid Green J (New Victoria Green, Brilliant Green). They are obtained when diethylaniline, $C_6H_5N(C_2H_5)_2$, is used instead of dimethylaniline, $C_6H_5N(CH_3)_2$. These colouring matters also occur in commerce as oxalates or sulphates. The oxalate, $C_{27}H_{32}N_2 + C_2H_2O_4 + H_2O$, loses its golden lustre by the action of the light. The zinc double salt has a red metallic lustre.

These colouring matters have a yellower tone than those obtained from dimethylaniline.

Helvetia-green, Acid Green.

These colouring matters are the sulphonic acids of the various sorts of benzaldehyde-greens. They are prepared by two methods : either the green base is treated with concentrated sulphuric acid, or the leuco-base is treated with sulphuric acid, and the resulting leuco-sulphonic acids are oxidised.

The monosulphonic acid of solid green, $C_{23}H_{23}N_2 \cdot SO_3H$, crystallises in green needles, with a reddish-brown reflex. It is easily soluble in hot water, with a green colour, but only sparingly soluble in cold water. The sodium salt is sparingly soluble in water. It forms white, silvery needles, which gradually become green in the air. The calcium, and most other salts, are sparingly soluble in water.

The sulphonic acid of solid green J is darker coloured, and possesses very little reflex.

Acid green dyes a brighter colour than solid green. It may be dyed from baths containing more acid, but is far less productive. Caustic soda solution gives no precipitate in dilute solutions (difference from solid green) ; strong soda ley gives a white precipitate. When dyed on the fibre, it gives nearly the same reactions as benzaldehyde-green. The behaviour towards concentrated hydrochloric acid serves to distinguish between the two. Helvetia-green is turned greenish-yellow, and the liquid is coloured yellow, while benzaldehyde-green turns a bright orange and gives up very little colour. Water reproduces the colour in both cases.

Light green S, Light green SF (yellow shade), *Acid Green SOF, Acid Green.*

This colouring matter, which is the sodium salt of the diethyldibenzyldiamidotriphenylcarbinol trisulphonic acid, is obtained by acting upon benzaldehyde with benzyl-ethyl aniline, $C_6H_5N(C_2H_5)(C_7H_7)$, sulphonating the resulting diethyldibenzyldiamidotriphenylmethane and oxidising the sulphonic acid thus obtained. Its constitution is represented by the formula:—

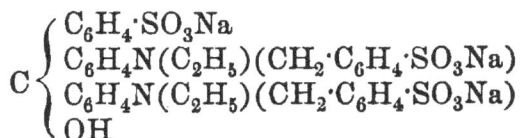

$$C \begin{cases} C_6H_4 \cdot SO_3Na \\ C_6H_4N(C_2H_5)(CH_2 \cdot C_6H_4 \cdot SO_3Na) \\ C_6H_4N(C_2H_5)(CH_2 \cdot C_6H_4 \cdot SO_3Na) \\ OH \end{cases}$$

The commercial product forms a light-green amorphous powder which dissolves in water with a green colour. The aqueous solution gives with hydrochloric acid a yellow-brown colouration, and a slight dirty violet precipitate with caustic soda.

Light green S dyes wool and silk a bright green in an acid bath, and is used largely, especially in wool-dyeing, in conjunction with other acid dyes for the production of compound shades. Thus, it may be used along with acid violet for the production of various shades of peacock blue.

Guinea green B is the sodium salt of the *di*sulphonic acid of diethyldibenzyldiamidotriphenylcarbinol, and forms a dark-green powder which resembles Light-green S in its properties and application.

Viridine, Alkali-green.

If in the preparation of malachite-green, the dimethyl-aniline is replaced by diphenylamine, $NH(C_6H_5)_2$, viridine is obtained. It probably possesses the constitution—

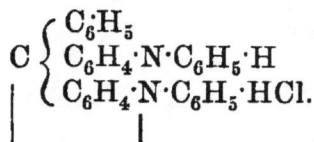

$$C \begin{cases} C_6H_5 \\ C_6H_4 \cdot N \cdot C_6H_5 \cdot H \\ C_6H_4 \cdot N \cdot C_6H_5 \cdot HCl. \end{cases}$$

It forms microscopic, bronze-coloured crystals, which are insoluble in water, but are soluble in alcohol with green colour.

By treating it with sulphuric acid, a sulphonic acid, insoluble in water, is formed. The soluble salts of the latter form the alkali-green of commerce.

A green colouring matter, probably identical with viridine, is formed by the oxidation of benzyldiphenyl-amine, $N(C_6H_5)_2C_7H_7$, with chloranil. It also yields an alkali-green with sulphuric acid.

Alkali-green only finds a very limited application, as it cannot be dyed in an acid bath. It is dyed in the same manner as alkali-blue, and may be used along with it for producing greenish-blue shades.

Aniline-red, Magenta.

Commercial analine-red is not a uniform substance, but a mixture of the salts of two bases—para-rosaniline, $C_{19}H_{19}N_3O$, and rosaniline, $C_{20}H_{21}N_3O$.

Since Hofmann's important investigations on aniline dyes, numerous attempts have been made to discover the constitution of the above bases. This question was first solved by MM. Emil and Otto Fischer.

One of the two bases, para-rosaniline, like benzaldehyde-green, is a derivative of triphenylmethane. The other, rosaniline, is derived from tolyldiphenymethan. They possess the formulæ—

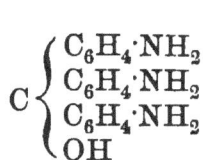

$$C \begin{cases} C_6H_4 \cdot NH_2 \\ C_6H_4 \cdot NH_2 \\ C_6H_4 \cdot NH_2 \\ OH \end{cases}$$

Para-rosaniline.

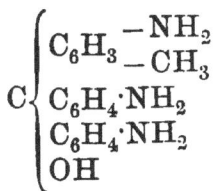

$$C \begin{cases} C_6H_3 {-NH_2 \atop -CH_3} \\ C_6H_4 \cdot NH_2 \\ C_6H_4 \cdot NH_2 \\ OH \end{cases}$$

Rosaniline.

They unite with acids with elimination of water. Thus, with hydrochloric acid, they form—

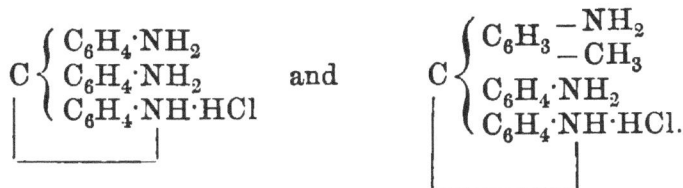

$$C \begin{cases} C_6H_4 \cdot NH_2 \\ C_6H_4 \cdot NH_2 \\ C_6H_4 \cdot NH \cdot HCl \end{cases} \quad \text{and} \quad C \begin{cases} C_6H_3 {-NH_2 \atop -CH_3} \\ C_6H_4 \cdot NH_2 \\ C_6H_4 \cdot NH \cdot HCl. \end{cases}$$

In order to give a clearer idea of the formulæ, a description of the synthesis of rosaniline will be given here.

(1.) If triphenylmethane is treated with nitric acid, trinitrotriphenylmethane is obtained.

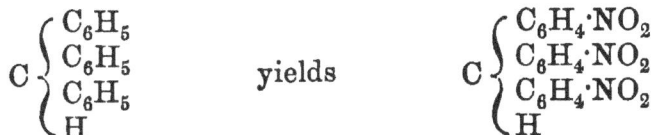

$$C \begin{cases} C_6H_5 \\ C_6H_5 \\ C_6H_5 \\ H \end{cases} \quad \text{yields} \quad C \begin{cases} C_6H_4 \cdot NO_2 \\ C_6H_4 \cdot NO_2 \\ C_6H_4 \cdot NO_2 \\ H \end{cases}$$

Triphenylmethane. Trinitrotriphenylmethane.

This, when treated with chromic acid, yields—

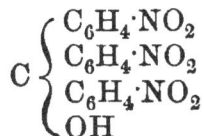

$$C \begin{cases} C_6H_4 \cdot NO_2 \\ C_6H_4 \cdot NO_2 \\ C_6H_4 \cdot NO_2 \\ OH \end{cases}$$

Trinitrotriphenylcarbinol

which, when partially reduced by zinc powder and acetic acid, is converted to para-rosaniline; *i.e.*, triamidotriphenylcarbinol.

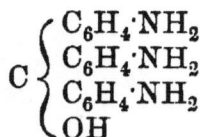

$$C \begin{cases} C_6H_4 \cdot NH_2 \\ C_6H_4 \cdot NH_2 \\ C_6H_4 \cdot NH_2 \\ OH \end{cases}$$

The para-rosaniline immediately unites with acetic acid, to form acetate of para-rosaniline.

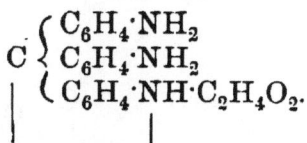

$$C \begin{cases} C_6H_4 \cdot NH_2 \\ C_6H_4 \cdot NH_2 \\ C_6H_4 \cdot NH \cdot C_2H_4O_2. \end{cases}$$

Rosaniline may be prepared from tolyldiphenylmethane in a similar manner.

(2.) A method, analogous to the preparation of benzaldehyde-green, has been patented, but is without practical value.

By heating the hydrochlorides of aniline, and para-amidobenzaldehyde with zinc chloride, para-leucaniline is obtained according to the equation :—

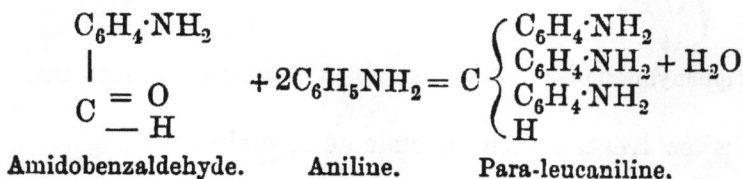

$$\begin{array}{c} C_6H_4 \cdot NH_2 \\ | \\ C = O \\ - H \end{array} \quad + 2C_6H_5NH_2 = C \begin{cases} C_6H_4 \cdot NH_2 \\ C_6H_4 \cdot NH_2 + H_2O \\ C_6H_4 \cdot NH_2 \\ H \end{cases}$$

Amidobenzaldehyde.   Aniline.   Para-leucaniline.

The para-leucaniline may be oxidised with chloranil, $C_6Cl_4O_2$, or manganese dioxide, to para-rosaniline.

(3.) Para-nitrobenzaldehyde acts on aniline sulphate according to the equation :—

$$\begin{matrix} C_6H_4 \cdot NO_2 \\ | \\ C {=O \atop -H} \end{matrix} + 2C_6H_5NH_2 = C \begin{cases} C_6H_4 \cdot NO_2 \\ C_6H_4 \cdot NH_2 \\ C_6H_4 \cdot NH_2 \\ H \end{cases} + H_2O$$

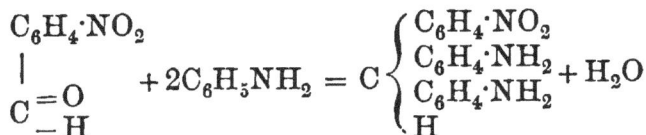

Nitrobenzaldehyde.   Aniline.   Nitrodiamidotriphenylmethane.

The purified product is heated with ferrous chloride, whereby the nitro group is reduced to an amido group, and the ferric chloride formed converts the methane hydrogen into hydroxyl, so that in one operation para-rosaniline is produced.

Rosaniline may be prepared by this method, if 1 molecule of nitro-benzaldehyde acts on 1 molecule of ortho-toluidine, $C_6H_4 {-CH_3 \atop -NH_2}$, and 1 molecule of aniline.

(4.) Further, by the action of benzaldehyde on aniline or a mixture of aniline and toluidine :—

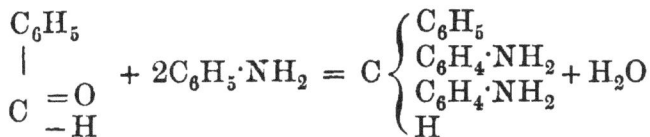

$$\begin{matrix} C_6H_5 \\ | \\ C {=O \atop -H} \end{matrix} + 2C_6H_5 \cdot NH_2 = C \begin{cases} C_6H_5 \\ C_6H_4 \cdot NH_2 \\ C_6H_4 \cdot NH_2 \\ H \end{cases} + H_2O$$

Benzaldehyde.   Aniline.   Diamidotriphenylmethane.

The diamidotriphenylmethane is treated with nitric acid, and the resulting nitrodiamidotriphenylmethane—

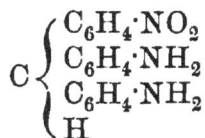

$$C \begin{cases} C_6H_4 \cdot NO_2 \\ C_6H_4 \cdot NH_2 \\ C_6H_4 \cdot NH_2 \\ H \end{cases}$$

is converted into para-rosaniline by method 3.

The formation of rosaniline from para-nitro, and from para-amido benzaldehyde, proves that at least one of the amido groups must be in the para position as regards the methane carbon. The study of the decomposition products of rosaniline has further shown the relative positions of the amido and methyl groups, so that the constitution may be expressed as follows :—

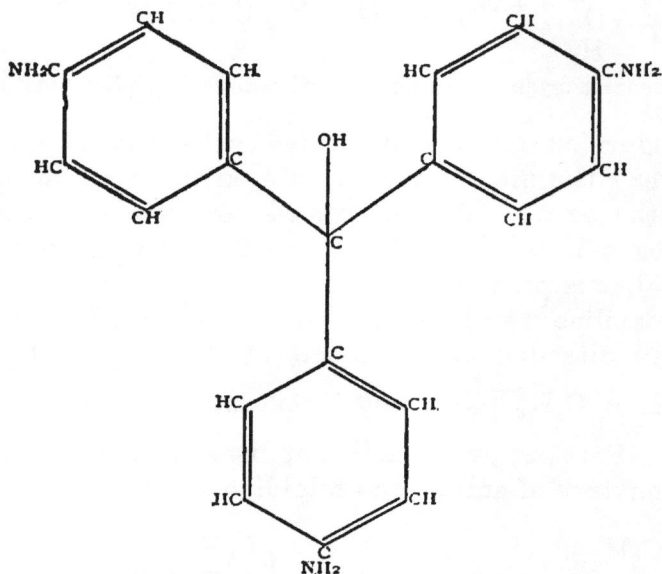

Para-rosaniline.

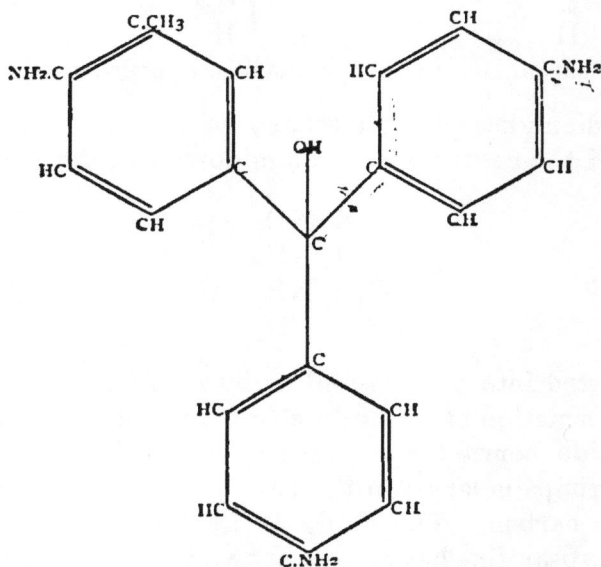

Rosaniline.

Commercial rosaniline is obtained by oxidation of aniline oil. From the constitution of rosaniline, it follows that para-toluidine

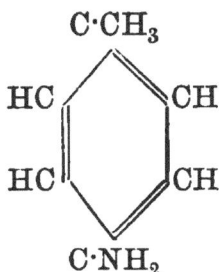

$$
\begin{array}{c}
\text{C·CH}_3 \\
\text{HC} \diagup\quad\diagdown \text{CH} \\
\text{HC} \quad\quad \text{CH} \\
\text{C·NH}_2
\end{array}
$$

must be contained in it, as it is the only toluidine which can yield the methane carbon in the para position, with regard to the amido groups.

The most beautiful reds are obtained when para-toluidine and aniline, or para-toluidine, aniline, and ortho-toluidine are oxidised according to the following equations :—

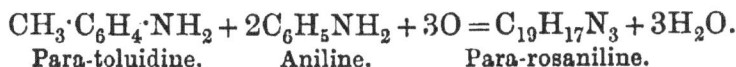

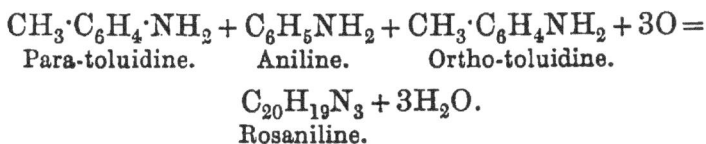

$$CH_3 \cdot C_6H_4 \cdot NH_2 + 2C_6H_5NH_2 + 3O = C_{19}H_{17}N_3 + 3H_2O.$$
Para-toluidine.   Aniline.   Para-rosaniline.

$$CH_3 \cdot C_6H_4 \cdot NH_2 + C_6H_5NH_2 + CH_3 \cdot C_6H_4NH_2 + 3O =$$
Para-toluidine.   Aniline.   Ortho-toluidine.

$$C_{20}H_{19}N_3 + 3H_2O.$$
Rosaniline.

Meta-toluidine only occurs in small quantities in commercial toluidine. It yields no red with para-toluidine.

The aniline oil containing aniline and toluidine in the proper proportions for aniline-red is known as " red oil," or aniline oil for red.

Manufacture of Magenta.

(1) *Arsenic acid method.*—100 pts. of aniline oil are heated with 125 pts. of 75 per cent. arsenic acid for eight hours, in a boiler fitted with distilling tube. The temperature must

be somewhat above the boiling-point of aniline (182°) Water, and a part of the aniline, distil over. The residue is then boiled with water, and filtered from the insoluble part. The solution contains arseniate and arsenite of rosaniline, a yellow bye-product (chrysaniline), besides excess of arsenic acid and resinous substances.

The residue consists of mauvaniline, violaniline, and some chrysaniline.

To the solution containing arsenic a large excess of common salt is added. A double decomposition takes place, rosaniline hydrochloride (magenta) and arseniate of soda being formed. The excess of salt dissolves, and the colouring matter, being sparingly soluble in salt solution, separates out. This method of separating colouring matters by salt is much used, and is technically known as "salting out."

The separated magenta is then crystallised from water, or dissolved again, salted out, and crystallised. The mother-liquors yield an impure magenta (cerise, geranium, etc.).

(2) *Nitrobenzene process.*—Instead of using arsenic acid as an oxidising agent, nitrobenzene and ferrous chloride are employed. The latter is oxidised by the nitrobenzene, and then oxidises the aniline, so that it acts as a transmitter of oxygen. 100 pts. of aniline oil are mixed with two-thirds of the quantity of hydrochloric acid necessary to saturate them, then with 50 pts. of nitrobenzene, and heated, with gradual addition of 3 to 5 pts. of iron filings. The magenta is prepared from the crude melt, as in the arsenic-acid process. The bye-products contain much *induline*, but no chrysaniline.

(3) *Mercury Process.**— By oxidation of aniline with mercuric nitrate, a very pure nitrate of rosaniline is obtained, which may be converted into the hydrochloride (rubine) by double decomposition with common salt.

* This process is at present obsolete.

Properties of the Rosaniline Salts.

Commercial rosaniline contains besides para-rosaniline ; but as the properties of the two are almost identical, we shall not distinguish between them here. The salts with one molecule of acid have a metallic-green appearance in reflected light; in transmitted light, in thin layers, they are red. The solutions are crimson, and are not fluorescent.

Hydrochloride of rosaniline, Magenta, $C_{20}H_{19}N_3HCl$, forms rhombic crystals, rather sparingly soluble in pure water, but more readily in acidified water and in alcohol. Amyl alcohol also takes it up. It absorbs moisture, on standing in the air. With concentrated hydrochloric acid it gives brown needles of the triacid salt, $C_{20}H_{19}N_3\cdot3HCl$, which dissolve with a brown colour. By pouring into excess of water, the original mono-acid salt is produced.

Caustic alkalies, ammonia, baryta, and lime decompose magenta solutions, and separate free rosaniline in a crystalline state. Pure, freshly prepared rosaniline is colourless.

Reducing agents, as zinc and acetic acid, stannous chloride, sulphurous acid, decolourise magenta, forming salts of leucaniline. Leucaniline is not reconverted into rosaniline by atmospheric oxygen ; the reduced solutions of rosaniline remain colourless. (Distinction from Magdala-red and safranine.)

Aldehyde and alcoholic shellac solution convert magenta into blue colouring matters.

Strong oxidising agents (chlorine, chloride of lime, permanganate of potash) decolourise magenta solutions. Limited oxidation produces a new colouring matter.

A yellowish-red colouring matter, aniline-scarlet, is obtained by the action of hydrogen peroxide or nitrate of lead on magenta. Chromic acid gives a brown colouring matter.

Thus rosaniline shows some resemblance to aniline in

its reactions with oxidising agents. The reason of this is, that the amido groups of aniline and toluidine are contained in it in an unaltered state.

Nitrate of rosaniline, $C_{20}H_{19}N_3 \cdot HNO_3$, is sometimes prepared on a small scale by the mercury process, and occurs in commerce as a very pure colouring matter known as azaleïne.

Acetate of rosaniline, $C_{20}H_{19}N_3 \cdot C_2H_4O_2$, is the most soluble rosaniline salt. It forms large green crystals, which after some time become brownish-red.

Chromate of rosaniline is a brick-red powder, nearly insoluble in water.

Dilute magenta solutions give a red precipitate with *tannic acid.*

Commercial magenta.—Pure magenta should consist only of the salts of rosaniline and para-rosaniline. Commercial magenta generally contains water, mineral impurities, resinous substances, and if prepared by the arsenic-acid process, more or less arsenic (up to 6·5 per cent.).

The purest qualities are the blue shades. The yellow ones contain a yellow colouring matter, phosphine. Magenta prepared by the nitrobenzene process contains no phosphine.

Aniline-red is sold under many different names, as magenta, azaleïne, roseïne, fuchsine, new red, rubine, etc.

Magenta-violet, fuchsine V, is a mixture of magenta and mauvaniline hydrochloride.

Cerise is a colouring matter containing magenta, which is used in dyeing browns. The mother-liquors after salting out still contain small quantities of magenta, phosphine, and a brown colouring matter. They are precipitated with milk of lime ; the precipitate is dissolved in acidulated water, and salted out. It forms an amorphous brown mass, with a vitreous fracture.

Cardinal, amaranth, are names applied to mixtures of

colouring matters of which the chief constituent is magenta.

Pure *para-rosaniline hydrochloride*, obtained by heating together nitrobenzene and para-nitrotoluene with aniline, para-toluidine, iron, and hydrochloric acid, is also a commercial product. The crystals are somewhat more compact than those of ordinary crystallised magenta.

Testing of magenta.—Good magentas are always well crystallised, and dissolve in pure water, without leaving any residue. The testing of magenta by comparative dye-trials, as well as the testing for adulteration and impurities, has already been referred to in the general part.

A solution of pure magenta is entirely decolourised by sulphurous acid, while impure samples remain yellow or brown.

For the qualitative estimation of arsenic in magenta made by the arsenic-acid process, Marsh's test is used.

In order to estimate arsenic quantitatively the magenta is fused with soda and nitre, dissolved in water, filtered, the arsenic precipitated with magnesia mixture, and the precipitate weighed as magnesium ammonium arseniate.

Application of magenta.—Silk and wool are dyed in neutral baths, silk may also be dyed in a bath containing boiled-off liquor. In brightening silk, the shades become slightly bluer.

In the preparation of magenta for printing, it must be remembered that by warming with many organic bodies, a violet is produced. Such substances are impure spirit, shellac, some gums, tannin, and fats.

Fibres dyed with magenta give up part of their colour to boiling water, while soap removes it completely.

In calico-printing, magenta is used for the production of a brown colour (magenta-puce). Suitably thickened magenta is printed, and passed through a hot bath containing bichromate of potash and sulphuric acid, a process which resembles the production of aniline-black on cotton yarn. Magenta-puce is also obtained, if, in the receipts for printing aniline-black, the aniline salt is replaced by magenta,—that is, a mixture of magenta, chlorate of soda, and copper sulphide are printed and steamed. This process is, however, not used on a large scale.

Detection on the fibre.—The decolourisation with acids (formation of triacid salts), with alkalies (separation of free rosaniline), and with sodium sulphide (reduction to leucaniline), are characteristic. When dyed along with vegetable colouring matters, the detection is easy. An alcoholic extract is made, which is described below.

Detection of magenta in food, etc.—Chemically pure magenta, and that which is prepared without arsenic acid, is not poisonous. The use of magenta for colouring articles of food, toys, and confectionery, is nevertheless forbidden in some countries.

In late years, magenta has been largely used for colouring wines, and many methods have been proposed for distinguishing between magenta and the natural colouring matter of wine, and also for the recognition of other artificial and natural colouring matters used for colouring wines.

Romei bases his process for the detection of magenta in coloured liquids, upon its solubility in amyl alcohol. 50 c.c. of the wine are warmed with 10 c.c. of lead acetate solution, sp. gr. 1·32, in order to remove the natural colouring matters. To the filtered solution, 1 drop of acetic acid and 10 c.c. of amyl alcohol are added, when the whole is well agitated. If the wine is pure, the amyl alcohol will remain colourless. If magenta is present, it is coloured red; rosolic acid yields a yellow, and orchil a red-violet. The layer of amyl alcohol is removed, and shaken with dilute ammonia. If a red-violet colour is formed, rosolic acid is present; a blue-violet colouration indicates orchil, and a decolourisation indicates magenta.

Brunner's method is to stir the warmed wine with stearic acid. On cooling, the stearic acid is violet in presence of magenta. *Falière's* method is also easy to carry out. The wine is treated with caustic baryta or ammonia, when the rosaniline is liberated. The liquid is then shaken with ether, and a piece of silk or wool placed in the ethereal solution. When the ether has evaporated, the red colour on the fibre is developed with acetic acid, and the fibre is tested for magenta in the usual manner.

From some red wines, magenta separates along with the natural red colouring matter (oenolin) on the sides of the vessel. *Cotton* collects this precipitate, treats it with water, adds ammonia, and shakes with ether, etc.

Preserved fruits and confectionery 'give red-coloured solutions with water, and the aqueous solution is tested according to the method adopted for the red wines.

For the detection of magenta in orchil and cudbear, for the adulteration of which it has been used extensively, *Rawson's* method (see *Journ. Soc. Dyers and Col.*, 1888, p. 68) is the most exact.

From 1 to 2 grms. of cudbear (or an equivalent amount of orchil liquor) are boiled with 50 c.c. of alcohol, and afterwards diluted with 100 c.c. of water ; 15 to 20 c.c. of a strong solution of basic acetate of lead (sp. gr. 1·25) are then added, followed after stirring by a similar quantity of strong ammonia. The mixture is filtered, and if the amount of magenta present is to be *estimated*, the precipitate is washed with a solution containing 1 part of ammonia, 5 parts of alcohol, and 10 parts of water ; otherwise the washing may be neglected. With pure cudbear the filtrate is quite colourless ; if magenta be present it is either colourless or pink, according to the amount of ammonia present in the solution. The liquid is then acidulated with acetic acid, when the presence or absence of magenta is at once made apparent ; in the case of pure cudbear or orchil the solution remains colourless, whereas, if a salt of rosaniline be present, the well-known colour of magenta is immediately developed. If further proof be wanting, a small piece of worsted yarn may be dyed in the solution and afterwards tested in the usual way with such reagents as hydrochloric acid, caustic soda, and a mixture of hydrochloric acid and stannous chloride.

By means of this method it is possible to detect with certainty 1 part of magenta in 100,000 parts of cudbear.

Acid Magenta, Magenta S, Rubine S.

A mixture of the trisulphonic acids of rosaniline and para-rosaniline is obtained by heating dried rosaniline with fuming sulphuric acid to 100°—170°. The product is poured into water, and neutralised with milk of lime. After filtering from the gypsum, the lime salts are decomposed with soda. The carbonate of lime is removed by filtration, and the soda salts are evaporated to dryness. The normal soda salt is hygroscopic, and is converted by hydrochloric acid into the acid salt.

The constitution of the para-rosaniline trisulphonic acid is—

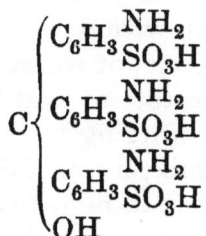

$$C\begin{cases} C_6H_3\,{}^{NH_2}_{SO_3H} \\ C_6H_3\,{}^{NH_2}_{SO_3H} \\ C_6H_3\,{}^{NH_2}_{SO_3H} \\ OH \end{cases}$$

Acid magenta forms a green metallic-looking powder, easily soluble in water, with a red colour. The solution is decolourised by alkalies, but no precipitate is formed. Dilute acids (also carbonic acid) reproduce the colour. The colour of the solution is only slightly altered by acids. Acid magenta is about half as strong, in dyeing, as ordinary magenta; but it can be dyed from strong acid baths, and may therefore be combined with other acid dyes and mordants. Thus it gives good results with indigo extract, Helvetia-green, and acid yellow.

On the fibre, acid magenta may be distinguished from magenta by its behaviour towards a mixture of equal parts of water and hydrochloric acid. Magenta is decolourised, acid magenta is unaltered, but the solution takes up some of the colour, and becomes cherry-red.

Magenta S has been used for colouring wines. It cannot be detected in the same manner as magenta, as it is insoluble in ether, and does not unite with stearic acid.

Aniline-blue.

(1.) *Spirit soluble Blue. Gentian-blue 6B, Spirit-blue O, Opal-blue, Bleu de nuit, Bleu lumière.*

If the hydrogen atoms of the amido groups of magenta are replaced by organic radicals, the colour becomes violet or blue. The shade is bluer, the more hydrogen atoms are replaced in this manner. The ethyl derivatives are the reddest, then come the salts of the methyl and benzyl rosaniline, and the purest blue is the hydrochloride of triphenylrosaniline.

Blue from rosaniline.—In order to prepare commercial spirit-blue, rosaniline (prepared from the purest blue shade of magenta by precipitation with ammonia or lime) is heated with a large excess of aniline (ten times the quantity) and some benzoic acid. The operation takes place in a distilling vessel. The temperature is the boiling-point of aniline, the excess of which distils over, along with ammonia water.

When a small test taken out shows that the action is finished, the whole is cooled to 60°, and saturated with hydrochloric acid. The blue separates, and the aniline hydrochloride dissolves. The precipitate is well washed with dilute hydrochloric acid, and then with water, dried and powdered, and sold as spirit-blue.

M

The formation of the blue base is shown in the following equation :—

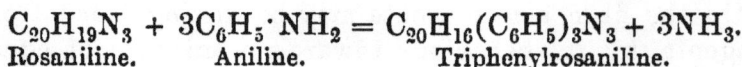

$$C_{20}H_{19}N_3 + 3C_6H_5 \cdot NH_2 = C_{20}H_{16}(C_6H_5)_3N_3 + 3NH_3.$$
Rosaniline.   Aniline.   Triphenylrosaniline.

The part which benzoic acid plays in the manufacture is not known ; but it is necessary, in order to obtain good results, to add a monobasic acid, such as stearic or benzoic acid.

Spirit-blue is further purified by dissolving in aniline, and precipitating with hydrochloric acid, or by other methods.   This finer product is known as basic-blue or opal-blue.

The less pure products have a reddish shade, which is especially noticeable in artificial light.   The best qualities give colours which appear pure blue by gas- or lamp-light, and are sometimes sold as light-blue (bleu lumière). They are also distinguished as spirit-blue 5B or 6B, the direct blue as B, while the intermediate products are known as 2B, 3B, 4B, according to their purity.

A blue with a decided red tone is obtained when magenta is used instead of rosaniline, and acetate of soda instead of benzoic acid, in the manufacture.   Spirit-blue R, or Parma-blue, is an intermediate product between the spirit-blues and the spirit-violets.

Spirit-soluble Diphenylamine-blue.

Aniline-blue may also be prepared from diphenylamine. Diphenylamine, $C_{12}H_{11}N$, is a crystalline body, possessing an agreeable smell.   It dissolves easily in alcohol and ether, sparingly in water.   It is a weak base.

In order to prepare it, equal molecules of aniline and aniline hydrochloride are heated to 250°.

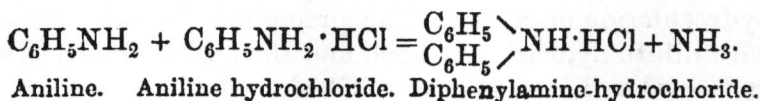

$$C_6H_5NH_2 + C_6H_5NH_2 \cdot HCl = \begin{matrix} C_6H_5 \\ C_6H_5 \end{matrix}\Big\rangle NH \cdot HCl + NH_3.$$
Aniline.   Aniline hydrochloride. Diphenylamine-hydrochloride.

Diphenylamine forms a blue colouring matter when heated with oxalic acid to 120°–130°.

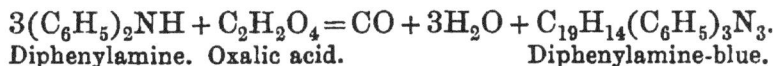

$$3(C_6H_5)_2NH + C_2H_2O_4 = CO + 3H_2O + C_{19}H_{14}(C_6H_5)_3N_3.$$
Diphenylamine. Oxalic acid.   Diphenylamine-blue.

The excess of oxalic acid is removed by washing with water, unaltered diphenylamine by boiling with benzene, when the residue is transformed into the hydrochloride and purified in the usual manner.

Diphenylamine-blue is of a finer quality, but at the same time more expensive than rosaniline-blue.

The chemical constitution of these two dye-stuffs is not identical. The rosaniline-blue is a derivative of commercial rosaniline, which consists chiefly of rosaniline, $C_{20}H_{19}N_3$. On the other hand, diphenylamine-blue is the hydrochloride of triphenylated para-rosaniline, and has the formula :—

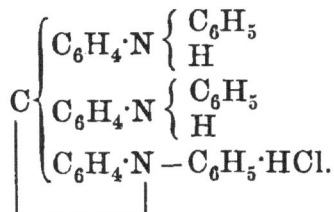

$$C \begin{cases} C_6H_4 \cdot N \begin{cases} C_6H_5 \\ H \end{cases} \\ C_6H_4 \cdot N \begin{cases} C_6H_5 \\ H \end{cases} \\ C_6H_4 \cdot N - C_6H_5 \cdot HCl. \end{cases}$$

Methyl- and ethyl-blue.—The above constitutional formula of diphenylamine-blue shows that it still contains two replaceable hydrogen atoms in the amido groups.

It has hitherto not been possible to introduce more than three phenyl groups, but the hydrogen atoms are easily replaceable by ethyl and methyl, and the colouring matters produced possess a still purer tone than diphenylamine-blue. In order to produce methyl- or ethyl-blue, methyldiphenylamine or ethyldiphenylamine—

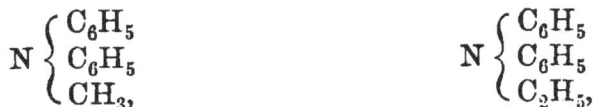

$$N \begin{cases} C_6H_5 \\ C_6H_5 \\ CH_3, \end{cases} \qquad N \begin{cases} C_6H_5 \\ C_6H_5 \\ C_2H_5, \end{cases}$$

is heated with oxalic acid. These bases are produced by the action of methyl or ethyl alcohol on diphenylamine hydrochloride :—

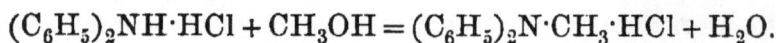

$$(C_6H_5)_2NH \cdot HCl + CH_3OH = (C_6H_5)_2N \cdot CH_3 \cdot HCl + H_2O.$$

Or the colouring matters may be prepared by the action of methyl or ethyl chloride on diphenylamine-blue.

The purest aniline-blue is obtained by melting methyl-diphenylamine on a water-bath, with chloranil ($C_6Cl_4O_2$), and then heating to 130°. The cooled mass is powdered, washed with hydrochloric acid, dissolved in alcohol and precipitated with water.

The chloranil used in this process is obtained by the action of hydrochloric acid and chlorate of potash on phenol. In the pure state it forms golden leaflets, insoluble in water. It sublimes at 150°. It is a tetrachlor derivative of quinon, $C_6H_4H_2$. Its oxidising property depends on the fact that it readily takes up hydrogen, like quinon itself, and is converted into tetrachlorohydroquinon :—

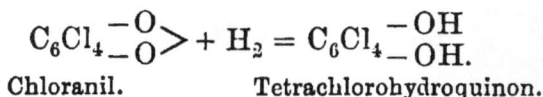

$$C_6Cl_4 {-O \atop -O} > + H_2 = C_6Cl_4 {-OH \atop -OH}.$$

Chloranil.   Tetrachlorohydroquinon.

Commercial chloranil contains besides trichloroquinon, which is, however, as powerful an oxidising agent as chloranil.

Properties of aniline-blue.—The hydrochloride of tri-phenylrosaniline is a brownish powder, which becomes pure blue at 100°. It is insoluble in water, but dissolves in boiling alcohol with a beautiful blue colour.

The acetate is more easily soluble in alcohol, and sometimes occurs in commerce in a state of solution.

Hydrochloric acid, nitric acid, and stannous chloride give blue precipitates in an alcoholic solution of aniline-blue. Caustic soda and ammonia give violet-blue precipitates, and at the boil, colourless solutions. If water is added to these solutions, free triphenylrosaniline, $C_{20}H_{17}(C_6H_5)_3N_3 \cdot OH$, is thrown down as a white precipitate, which rapidly becomes blue in the air.

Application.—The greatest part of the spirit-blue made is used in the manufacture of soluble-blue and alkali-blue.

The finest qualities, as spirit-blue 5B and 6B, diphenylamine-blue, methyl- and ethyl-blue, are used in silk-dyeing for producing light and very pure shades, which cannot be obtained with soluble blue. The colouring matter is dissolved in 40 to 50 parts of methylated spirits, filtered, and dyed in a bath containing boiled-off liquor.

In wool-dyeing, spirit-blue has been used for producing bright blues which have to stand milling; but it has been superseded for this purpose by Victoria-blue and night-blue, which yield faster colours. It is dyed in a bath containing alum or sulphuric acid. Brighter shades can be obtained by the addition of a little stannic chloride. On cotton it is best fixed by means of an alumina mordant. The material is first worked in a strong soap solution, well wrung and passed into a bath of alum, when oleate or stearate of alumina will be fixed on the fibre. The dye-bath is made up as for wool.

(2.) *Soluble-blues.*

Spirit-blue, on being heated with sulphuric acid, is easily transformed into sulphonic acids, some of which are soluble in water in the free state, others as their alkali salts.

Diphenylamine-blue, soluble in water, may be obtained direct by heating diphenylamine sulphonic acid—

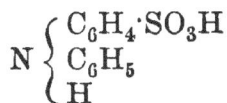

$$N \begin{cases} C_6H_4 \cdot SO_3H \\ C_6H_5 \\ H \end{cases}$$

with oxalic acid.

The first blue soluble in water was prepared by Nicholson in 1862, and was called, after him, "Nicholson's Blue."

The more sulpho groups are introduced into triphenylrosaniline, the more easily soluble the products become; but their fastness decreases in the same proportion, both as regards the action of light and air, soap and alkalies. The higher sulphonic acids, such as rosaniline tetrasulphonic acid, $C_{38}H_{27}N_3(SO_3H)_4$, are therefore never prepared.

The sulphonic acids are distinguished as follows :—

The monosulphonic acid is insoluble in pure or acidified water; its alkali salts dissolve with a light-brown colour, the solution becomes red-brown, when heated with an excess of alkali.

The disulphonic acid is soluble in water, but insoluble in water containing sulphuric acid. The solution in excess of alkali is yellow.

The trisulphonic acid dissolves both in pure water and in water containing sulphuric acid. Its alkaline solution is colourless with excess of alkali.

Alkali-blue.   Nicholson's blue.

Alkali-blue is the soda-salt of the monosulphonic acid, and is represented by the formula—

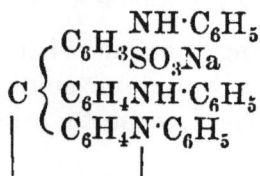

$$C \begin{cases} C_6H_3^{NH\cdot C_6H_5}_{SO_3Na} \\ C_6H_4NH\cdot C_6H_5 \\ C_6H_4N\cdot C_6H_5 \end{cases}$$

It is obtained by treating spirit-blue with sulphuric acid at 30° to 35° C., and pouring the resulting brown-yellow solution into water. The precipitate is collected, washed, and dissolved in the required (calculated) quantity of soda. The dye-stuff is then salted out, or evaporated to dryness, with addition of a little carbonate of ammonia.

Alkali-blue D is obtained by the action of strong sulphuric acid on diphenylamine blue, and is probably the sodium salt of the corresponding monosulphonic acid, $C_{37}H_{28}N_3SO_3Na$, while *Bavarian-blue DSF* is the sodium salt of the disulphonic acid with some of the trisulphonic acid.

Properties.—Alkali-blue comes into commerce as a brownish powder, or in lumps. It should dissolve without residue in about 5 parts of water.

By acidifying with acetic acid, the liquid is coloured blue; and on boiling, the free sulphonic acid separates as a blue precipitate.

By acidifying with hydrochloric acid, the free acid is completely precipitated; and the solution is rendered colourless when pure alkali-blue is present; but if the di- and tri-sulphonic acids are present, the solution remains coloured. If the test sample gives off carbonic acid on acidifying, it most likely contains soda.

Concentrated soda solution colours the solution red-violet; on boiling, red-brown. Excess of ammonia decolourises the solution. Stannous chloride produces a blue precipitate.

Application.—Alkali-blue is generally dyed in a feebly alkaline solution. It is not used in cotton-dyeing, as it will not combine with acid mordants; but it is used very extensively for bright-blue shades on silk and wool.

The dye-bath should be made perfectly free from lime-salts, as the lime-salt of the sulphonic acid is insoluble in water. Calcareous water can be made suitable by boiling with a small quantity of tin-salt. The dyeing is effected nearly at a boil, borax, soluble glass, soda or stannate of soda being added to the dye-bath, to render the dye more even and faster.

Alkali-blue is taken up by the fibre in an almost colourless condition. In order to develop the colour, the material is passed after dyeing through a weak, hot acid

bath ("developing"). The dyes are made faster to milling by employing alumina or tin-salts as mordants. Alkali
blue is little used for producing mixed colours, as it requires this peculiar treatment in dyeing, which is essentially different to the application of most other colouring
matters. Mixed shades can, however, be produced by adding various acid colours to the developing bath.

Soluble-blue, Water-blue, Cotton-blue.

Spirit-blue is heated with 3 to 4 parts of sulphuric acid
to 60° for a length of time, and finally the temperature is
raised to 100°–110°. After cooling, the product is mixed
with three or four times the quantity of water, to precipitate the colouring matter. It is then filtered off,
dissolved in a large quantity of boiling water, and the
excess of sulphuric acid removed by cautious addition of
milk of lime. The calcium sulphate is filtered off, and
the filtrate mixed with soda or carbonate of ammonia,
and evaporated to dryness.

Soluble-blue contains principally salts of triphenyl-
rosaniline- and triphenylpararosaniline- trisulphonic acid,
$C_{38}H_{28}N_3(SO_3H)_3$ and $C_{37}H_{26}N_3(SO_3H)_3$. The ammonia-
salt forms a mass with a coppery lustre, or a granular
powder; the soda-salt generally occurs as dark-blue
irregular lumps.

China-blue is a name given to a very porous water-blue,
which is obtained by adding to a very concentrated and
slightly acidulated solution of water-blue, carbonate of
ammonia. Methyl-blue, Methyl-blue BI. for cotton, Brilliant
cotton-blue greenish, methyl water-blue is the sodium salt of
pure triphenylpararosaniline trisulphonic acid, $C_{37}H_{26}N_3$
$(SO_3Na)_3$. The reactions of water-blue are similar to those
of alkali-blue. The colouring matter is not precipitated
by acids. Caustic soda decolourises the solution.

Application.—Soluble-blue is principally used in cotton dyeing. It is either fixed by means of tannin, along with alum, tartar emetic, or tin-salts. For some purposes it suffices to treat the cotton in a cold solution of the dye to which alum has been added.

This dye-stuff further serves for producing compound colours on wool and silk; it possesses the advantage over alkali-blue that it may be dyed from acid-baths, and may thus be combined with many other colouring matters. As a self-colour on wool and silk it is never used, as it is not so fast and not so productive as alkali-blue.

Detection of aniline-blue on the fibre.—Concentrated sulphuric acid decolourises. Water-blue gives a blue solution. Hydrochloric acid nearly decolourises spirit-blue and alkali-blue.

Spirit-blue first becomes light-blue and is then gradually decolourised by ammonia, while alkali- and water-blue disappear immediately.

Caustic soda turns spirit-blue to a grey-violet; alkali-blue to a yellow-brown; and water-blue to a red-brown.

Stannous chloride and hydrochloric acid decolourise spirit-blue and alkali-blue gradually, and remove water-blue with a pure blue colour.

Alcohol strips spirit- and alkali-blue even in the cold, while water-blue is not taken up even at a boil.

Phenyl-violet. Spirit-violet.

If the manufacture of aniline-blue is altered, so that less aniline is used, and the mixture heated for a shorter period, a mixture of mono- and a diphenylrosaniline is obtained, which contains very little triphenylrosaniline, and which, when combined with hydrochloric acid, finds a limited application as "spirit-violet."

It does not give as bright colours as methyl-violet, but they stand milling better, and therefore find some application in wool-dyeing.

Detection on the fibre.—Alcohol removes the colour, ammonia decolourises, caustic soda turns brown. Its behaviour towards stannous chloride serves to distinguish it from other violets. It is removed with a blue colour, and is only decolourised very slowly.

Hofmann's Violet. Dahlia, Primula. Red-violet, 5 R extra, Violet 5 R.

The Hofmann's violets consist of various ethyl and methyl derivatives of rosaniline.

For their preparation, an alcoholic solution of rosaniline is treated with methyl or ethyl iodide and caustic soda.

The quantity of iodide used depends upon the shade required; for red shades it is least, and for pure violet greatest. At most three methyl or ethyl groups may be introduced in this manner, according to the equation—

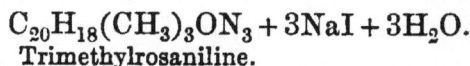

$$C_{20}H_{21}N_3O + 3CH_3I + 3NaHo =$$

Rosaniline.

$$C_{20}H_{18}(CH_3)_3ON_3 + 3NaI + 3H_2O.$$

Trimethylrosaniline.

The ethyl derivatives are redder than those of methyl.

The salt used in dyeing is the hydrochloride. It yields, when carefully prepared, a very pure, reddish violet.

The application of this dye is very limited, as it has been almost completely replaced by methyl-violet.

Methyl-violet. Paris-violet.

Preparation.—This violet is produced by direct oxidation of the purest dimethylaniline, $C_6H_5N(CH_3)_2$, (free from methyl toluidine), with copper chloride.

The copper chloride is prepared by double decomposition of nitrate or sulphate of copper with common salt. The solution is mixed with a large quantity of common salt, and dimethylaniline and acetic acid are poured in. The mass is moulded into cakes, and dried at 40° to 50°, when it assumes a green metallic appearance. It is then extracted with a quantity of boiling water, insufficient to dissolve all the salt. The violet is insoluble in the salt solution, and therefore remains behind.

The residue is dissolved in water and the copper precipitated by sulphuretted hydrogen. The copper sulphide is removed by filtration, when the colouring matter is salted out, and is purified by re-dissolving in water and salting out again; the product is then either crystallised or converted into the zinc double salt.

In place of the large quantity of salt used to modify the action of the copper chloride on dimethylaniline, sand may be conveniently substituted. To remove the copper, gaseous sulphuretted hydrogen, or sodium sulphide, is used.

A colouring matter known as *chloranil-violet*, which is probably identical with methyl-violet, is obtained by the oxidation of dimethylaniline with chloranil. (See p. 164.)

The base contained in methyl-violet is pentamethyl-pararosaniline—

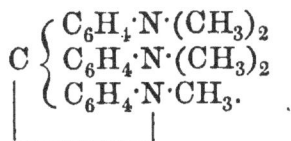

$$C \begin{cases} C_6H_4 \cdot N \cdot (CH_3)_2 \\ C_6H_4 \cdot N \cdot (CH_3)_2 \\ C_6H_4 \cdot N \cdot CH_3. \end{cases}$$

Its formation is expressed by Caro and Graebe in the equation—

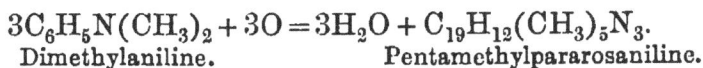

$$3C_6H_5N(CH_3)_2 + 3O = 3H_2O + C_{19}H_{12}(CH_3)_5N_3.$$
Dimethylaniline.   Pentamethylpararosaniline.

It is distinguished from Hofmann's violet in its chemical constitution, being a pentamethylated pararosaniline, while Hofmann's violet is only trimethylated rosaniline.

Properties.—Methyl-violet (mark B) comes into commerce as a hydrochloride, $C_{19}H_{12}(CH_3)_5N_3 \cdot HCl$, or as a
zinc double salt.   The latter forms small crystals, while
the first either consists of a powder, or of irregular lumps.
Methyl-violet has a green metallic reflex; it is easily
soluble in water and alcohol.

Dilute solutions are turned pure blue by very little
hydrochloric acid; more acid makes them dichroïc.   Thin
layers are then green, while thicker ones are much less
transparent, and of a red colour.   An excess of acid turns
the solution yellow, owing to the formation of acid salts.
Ammonia produces a lilac-coloured precipitate, and caustic
soda a brown-violet one, the solution becoming colourless
on boiling.

Chromic acid gives a dirty violet precipitate, chloride
of lime decolourises, and stannous chloride gives a blue-
violet precipitate, becoming lighter on boiling.

For dyeing, 1 part of colour is dissolved in 50 to 100
parts of water, and the solution filtered.   Silk is dyed in
a bath of acidulated "boiled-off liquor."   The shades
may be brightened by tartaric acid or very little sulphuric
acid.   Cotton is dyed by the tannin and tartar-emetic
process.

Methyl-violet is also used in calico-printing.   It is fixed
by albumen, glycerin and arsenic or some form of tannin.
It is also used for topping goods dyed with iron mordants
and alizarin, in order to brighten and beautify the fast
violet produced.   Its aqueous solution thickened with
glycerin is also used for marking with india-rubber
stamps.

Detection on the fibre.—Methyl-violet is gradually removed from the textile fibres by boiling with water.
Hydrochloric acid removes part of the colour; the fibre
is coloured greenish yellow, but the original colour is
restored on washing with water.   Ammonia decolourises,
caustic soda turns it red-violet and gradually decolourises

it. A mixture of stannous chloride and hydrochloric acid gives a greenish yellow to yellow colour.

Regina-purple.

This colouring matter consists of the acetate of orthotolylpararosaniline, and is obtained by acting upon the product which distils over in the manufacture of magenta by the arsenic acid process with magenta base and acetic acid at 120°. It forms a green powder which dyes a red shade of violet.

Benzyl-rosaniline-violet.

(Methyl-violet 6B.)

On heating methyl-violet with benzyl chloride, $C_6H_5CH_2Cl$, alcohol, and soda or lime, in a vessel provided with an upright condenser, part of the methyl groups are replaced by benzyl, C_7H_7, and a series of bases is formed, of which the first members are—

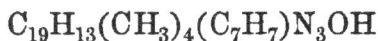

$$C_{19}H_{13}(CH_3)_4(C_7H_7)N_3OH$$
and
$$C_{19}H_{13}(CH_3)_3(C_7H_7)_2N_3OH.$$

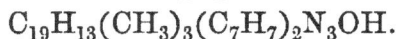

The excess of alcohol and benzyl chloride is distilled off, and the colouring matter is purified in the usual manner, by salting out, etc.

It comes into commerce as the hydrochloride or the zinc double salt. The highest benzylated product is marked 6B, and between it and methyl-violet are 2B, 3B, 4B, 5B, produced by mixing the brand B and 6B.

Benzyl-violet is very similar in its reactions to methyl-violet. It gives somewhat bluer colours.

Detection on the fibre.—Treated with caustic potash ley, methyl-violet becomes red-violet; benzyl-violet, light-blue. Both are decolourised on standing.

Crystal-violet. Violet 6B.

This colouring matter is the hydrochloride of hexa-methylpararosaniline—

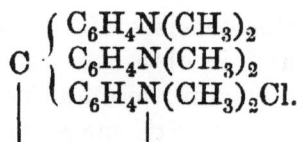

$$C \begin{cases} C_6H_4N(CH_3)_2 \\ C_6H_4N(CH_3)_2 \\ C_6H_4N(CH_3)_2Cl. \end{cases}$$

It may be obtained by the action of perchlormethyl formiate or dimethylaniline in presence of aluminium chloride. It is also formed, according to another method, by the action of carbon oxychloride $COCl_2$ on dimethyl-aniline :—

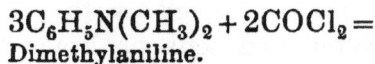

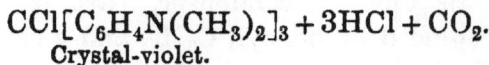

$$3C_6H_5N(CH_3)_2 + 2COCl_2 =$$
Dimethylaniline.

$$CCl[C_6H_4N(CH_3)_2]_3 + 3HCl + CO_2.$$
Crystal-violet.

It occurs in the form of well-defined crystals which possess a peculiar greenish brown metallic reflex. They dissolve in water and alcohol with a deep violet-blue colour, and crystallise easily from the former, but not from the latter solution. The leuco-base has the formula $C_{25}H_{21}N_3$ and melts at 173°.

The colouring matter yields in dyeing a very blue shade of violet; otherwise it resembles ordinary methyl-violet, not only in its application, but also in its reactions

Ethyl-violet.

This colouring matter is formed by the action of di-ethylaniline on tetraethyldiamidobenzophenon chloride. It is the hydrochloride of hexaethylpararosaniline, $C_{19}H_{12}N_3$ $(C_2H_5)_6Cl.$

The commercial product forms a green crystalline powder easily soluble in water with a violet colour. It resembles methyl-violet in its properties and application.

Acid-violet, 6B.

This colouring matter consists essentially of the sodium salt of pentamethylbenzylpararosanilinemonosulphonic acid—

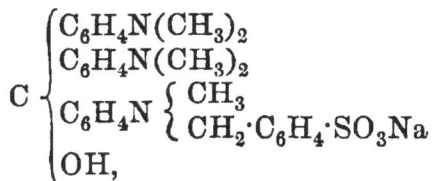

$$C \begin{cases} C_6H_4N(CH_3)_2 \\ C_6H_4N(CH_3)_2 \\ C_6H_4N \begin{cases} CH_3 \\ CH_2 \cdot C_6H_4 \cdot SO_3Na \end{cases} \\ OH, \end{cases}$$

and is obtained by the oxidation of the corresponding leuco-compound.

The commercial product forms a dark-violet powder, soluble in water with a violet, in alcohol with a blue colour. It dyes wool and silk violet in an acid bath, and is largely used along with other acid colours for the production of mixed shades.

Red-violet 4 *R.S.* is the sodium salt of dimethylrosanilinetrisulphonic acid, and is formed by the action of fuming sulphuric acid on dimethylrosaniline. It forms a reddish-violet powder which dyes wool and silk in an acid bath. The shades obtained are somewhat bluer than those obtained with acid magenta.

Methyl-green.

Preparation.—Green colouring matters are formed when methyl- and ethyl-rosanilines are heated with excess of methyl chloride, or iodide, or ethyl chloride and iodide.

A similar dye was formerly prepared by the action of methyl iodide on Hofmann's violet or rosaniline, and was known as iodine-green, bright-green, or vert lumière. It is now replaced by methyl-green, which is prepared from methyl chloride and methyl-violet. Methyl-violet is

dissolved in alcohol, and the free base precipitated in a
fine state of division, by caustic soda or baryta.   The base
is mixed with methyl-chloride, and heated some hours to
60°–70° in horizontal, closed cylinders.   The mass is
poured into water, the unaltered violet base* filtered off,
and the filtrate neutralised with hydrochloric acid.   By
the addition of a very small quantity of salt solution a
further quantity of methyl-violet is precipitated. The green
dye is then precipitated with salt or zinc chloride, and the
precipitate purified by boiling with alcohol.   Methyl-
green is formed by the addition of methyl-chloride to
methyl-violet, as is expressed by the equation—

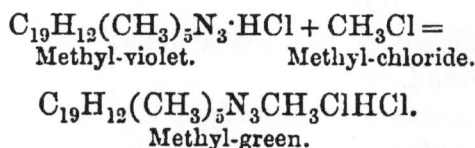

$$C_{19}H_{12}(CH_3)_5N_3 \cdot HCl + CH_3Cl =$$

Methyl-violet.   Methyl-chloride.

$$C_{19}H_{12}(CH_3)_5N_3CH_3ClHCl.$$

Methyl-green.

Accordingly, it possesses the constitution—

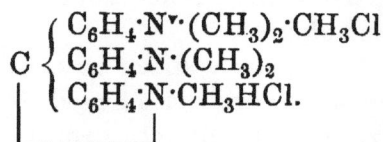

$$C \begin{cases} C_6H_4 \cdot N^{\cdot\cdot} \cdot (CH_3)_2 \cdot CH_3Cl \\ C_6H_4 \cdot N \cdot (CH_3)_2 \\ C_6H_4 \cdot N \cdot CH_3HCl. \end{cases}$$

Properties.—Methyl-green comes into commerce as the
zinc double salt, in small crystals.   The salt—

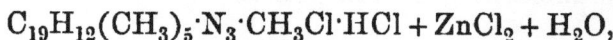

$$C_{19}H_{12}(CH_3)_5 \cdot N_3 \cdot CH_3Cl \cdot HCl + ZnCl_2 + H_2O,$$

forms green needles; a salt containing more zinc forms
large coppery prisms.   Another variety occurs in the form
of a light-green powder.

All the varieties dissolve in water and alcohol, with a
green colour, but they are insoluble in amyl alcohol.
(Distinction from benzaldehyde-green.)

Hydrochloric acid colours the solution greenish yellow,

* The green base is soluble in water.

a triacid salt being formed. On diluting with water, the original colour is produced. Stannous chloride gradually decolourises, and chloride of lime destroys the colour.

The behaviour of methyl-green towards alkalies requires a short explanation. The constitutional formula shows it to be the chlormethylate of a tertiary base, containing a pentatomic nitrogen atom, N^v, which is combined with phenylene, C_6H_4, three methyl groups, CH_3, and one chlorine. The simplest example of this class of bodies is tetramethylammonium chloride—

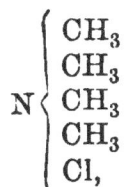

$$N \begin{cases} CH_3 \\ CH_3 \\ CH_3 \\ CH_3 \\ Cl, \end{cases}$$

which is not decomposed by alkalies, but which gives up its chlorine to moist silver oxide. A strong base, soluble in water, is thus formed :—

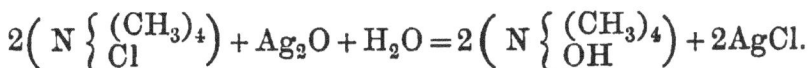

$$2\left(N \begin{Bmatrix} (CH_3)_4 \\ Cl \end{Bmatrix} \right) + Ag_2O + H_2O = 2\left(N \begin{Bmatrix} (CH_3)_4 \\ OH \end{Bmatrix} \right) + 2AgCl.$$

Tetramethylammonium   Tetramethylammonium
chloride.   hydrate.

By decomposing methyl-green solution with dilute alkalies, a reddish solution is produced, which on dilution becomes colourless. Concentrated alkalies produce a resinous precipitate, containing chlorine, which is soluble in pure water. This compound is the base corresponding to rosaniline :—

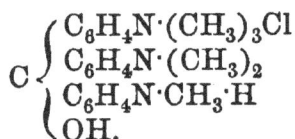

$$C \begin{cases} C_6H_4N \cdot (CH_3)_3Cl \\ C_6H_4N \cdot (CH_3)_2 \\ C_6H_4N \cdot CH_3 \cdot H \\ OH. \end{cases}$$

When its aqueous solution is treated with silver oxide,

the chlorine is removed, and the easily soluble ammonium base is produced :—

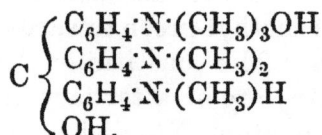

$$C \begin{cases} C_6H_4 \cdot N \cdot (CH_3)_3OH \\ C_6H_4 \cdot N \cdot (CH_3)_2 \\ C_6H_4 \cdot N \cdot (CH_3)H \\ OH. \end{cases}$$

By heating methyl-green to 100°, methyl chloride is gradually given off, the decomposition being very rapid at 130°.   Methyl-violet remains behind :—

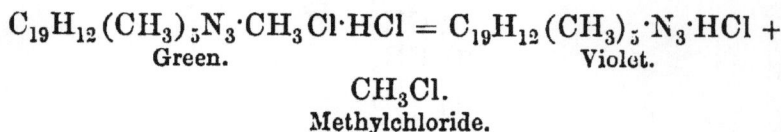

$$C_{19}H_{12}(CH_3)_5N_3 \cdot CH_3Cl \cdot HCl = C_{19}H_{12}(CH_3)_5 \cdot N_3 \cdot HCl +$$
Green.   Violet.

$$CH_3Cl.$$
Methylchloride.

Picric acid, $C_6H_2(NO_2)_3OH$, produces in solutions of methyl-green a dark-green, crystalline precipitate, which contains no chlorine.

The picrate of methyl-green is nearly insoluble in water, but soluble in alcohol.   It is sometimes sold as " spirit-soluble green."   Methyl-green is easily distinguished from other dye-stuffs by the above reactions.

In order to test the value of methyl-green, a dye-trial and an estimation of the ash are made.   The sample should be entirely soluble in water, with a blue-green colour.   If a green residue remains, it may consist of the picrate, which, however, dissolves completely in alcohol or dilute soda solution.

If the picrate contains excess of picric acid, this is easily detected by a dye-trial.   A skein of silk or wool is placed in the solution, and removed.   A second skein placed in the same solution is dyed yellow, if excess of picric acid is present.

If the aqueous solution of methyl-green gives a precipitate with alkalies, the presence of methyl-violet is indicated; this is owing to defective purification. The precipitate is filtered off, and the colour of its solution in hydrochloric acid is noticed.

An adulteration with soluble blue is recognised by adding a saturated solution of picric acid. The green is completely precipitated, and the liquid acquires a greenish tinge in presence of aniline-blue. In extreme cases it may be bluish-green, indicating the presence of large quantities of aniline-blue.

Methyl-green should be entirely soluble in boiling alcohol. Frequently a residue remains of a greenish-white powder, which is also, like methyl-green, turned violet on being heated to 130°. It is the chloride of nona-methyl para-leukaniline :—

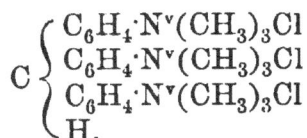

$$C \begin{cases} C_6H_4 \cdot N^v(CH_3)_3Cl \\ C_6H_4 \cdot N^v(CH_3)_3Cl \\ C_6H_4 \cdot N^v(CH_3)_3Cl \\ H. \end{cases}$$

It is often formed, as a bye-product, in considerable quantities. In purifying methyl-green with alcohol, on a large scale, it is left behind; and when dried and powdered, often serves for adulterating methyl-green. It is easily soluble in water, and is therefore not detected in the ordinary use of the dye-stuff.

Application.—In dyeing with methyl-green, the dye-bath should never reach the boil, as some violet is likely to be produced, and this would influence the desired colour. This decomposition takes place still more readily in presence of mineral acids. For the same reason, goods printed with methyl-green should not be steamed for a long time.

Silk is dyed in a lukewarm bath containing boiled-off liquor. In brightening, tartaric or acetic acid is used. If a yellower tone is required, picric acid is added to the brightening bath.

An aqueous, warm solution of methyl-green does not dye wool well, only bluish-grey shades being produced. Better results are obtained when the dyeing is done in an alkaline bath, and the colour is subsequently developed with acid. Another method is to fix with tannin.

But the best results are obtained when the wool is mordanted with sulphur, according to Lauth's process. For this purpose, the wool is first mordanted with hyposulphite (thiosulphate) of soda and hydrochloric acid, alum or sulphuric acid. It is then well washed and dyed with methyl-green. Acetate of zinc is frequently added to the dye-bath, and this is especially necessary when picric acid has been added for the production of a yellower shade. The zinc salt is gradually decomposed by the sulphur contained in the wool, zinc sulphide being formed; and this is as good a mordant for methyl-green as free sulphur. A little of the methyl-green is taken up by the fibre, the acetic acid produced by the action of the sulphur on the zinc acetate fixes a quantity of picric acid, and then a little acetate of soda is added, and the methyl-green is all fixed. For wool-dyeing, methyl-green has been almost entirely replaced by acid-green and benzaldehyde-green, as these dyes require no mordant, are not altered by heat, and they serve better for compound colours. Cotton is mordanted by the tannin process.

Detection on the fibre.—The reaction on heating is very characteristic for methyl-green. A bit of the material is heated a little above 100°, when it becomes violet if dyed with methyl-green.

The following reactions also serve for its detection :— Excess of hydrochloric acid colours the fibre yellow ; on washing with water, the original colour is reproduced. Stannous chloride, ammonia, and soda-ley decolourise. Acetic acid or alcohol remove the colour, yielding blue-green solutions.

Colouring matters derived from tetramethyldiamidobenzo-phenon.

These colouring matters are obtained by the action of tetramethyldiamidobenzophenon,

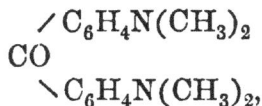

$$CO \Big\langle \begin{array}{l} C_6H_4N(CH_3)_2 \\ C_6H_4N(CH_3)_2, \end{array}$$

on the hydrochlorides of certain amines in presence of condensing agents. They are all basic colouring matters, and are used chiefly in the dyeing of cotton.

Auramines.—Ordinary auramine is obtained in the following manner: 25 kilos. tetramethyldiamidobenzophenon are well mixed with 25 kilos. ammonium chloride and 25 kilos. zinc chloride, and the mixture is heated for 4–5 hours to 150°–160° C., care being taken to agitate it well from time to time. The reaction is over as soon as a sample taken out dissolves in water. The melt is then allowed to cool, when it is broken up and extracted first with cold water, acidulated with hydrochloric acid in order to remove most of the unchanged ammonium chloride and the zinc chloride. It is then extracted with boiling water, and the colouring matter is precipitated from the filtered solution by the addition of common salt. The crystalline precipitate obtained in this way can be further purified by crystallising from water.

Auramine is the hydrochloride of imidotetramethyldi-amidodiphenylmethane, a colourless base which forms with acids intensely yellow and for the most part well crystallised salts. It possesses the formula—

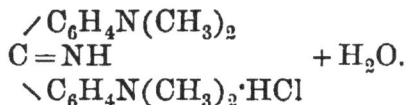

$$C=NH \Big\langle \begin{array}{l} C_6H_4N(CH_3)_2 \\ C_6H_4N(CH_3)_2 \cdot HCl \end{array} \quad + H_2O.$$

The hydrochloride, sulphate, and acetate are easily soluble in water, while the double compound with zinc chloride, as well as the sulphocyanide, are only very sparingly soluble in cold water. Mineral acids, when added to the solution, produce at first no change, but if allowed to act for a length of time, or if boiled with them, the solution is decolourised, the colouring matter being reconverted into the ketonbase and ammonia. Alkaline reducing agents, such as sodium amalgam, gradually decolourise the alcoholic solution. On adding water, a colourless crystalline reduction product is thrown down, which, when treated with acetic acid and heated, gives a deep blue colouration. If auramine is heated, with aniline up to the boiling point of the latter, the mixture evolves ammonia and becomes orange, phenylauramine being formed.

The auramine of commerce is a yellow powder which dissolves easily in water with a yellow colour. It is fixed on cotton with tannin and tartar emetic, and yields an extremely pure and brilliant yellow, which is sufficiently fast to soap and light for ordinary purposes. It can also be dyed on cotton mordanted with Turkey-red oil. The temperature in dyeing should not exceed 70° C., since at higher temperatures the colouring matter is dissociated with re-formation of ketonbase and ammonium chloride.

Detection on the fibre.—The fibre is decolourised by strong sulphuric or hydrochloric acid, as well as by caustic potash solution.

If, in the preparation of auramine according to the above method, the ammonium chloride is replaced by the hydrochlorides of aniline, toluidine, naphthylamine, or other aromatic amines, substituted auramines are formed, which vary in shade from yellow to light-brown. Thus metaxylylauramine yields in dyeing golden-yellow shades, phenyl- and paratolyl-auramines yield orange shades, while the metaphenylenediamine derivative yields an orange-brown.

Victoria blue B.—If tetramethyldiamidobenzophenon is treated with phenyl-α-naphthylamine and phosphorus oxychloride, a blue melt is obtained. The melt is first extracted with cold water and is then dried, in which state it comes into the market as Victoria blue B.

Victoria blue B possesses the constitutional formula—

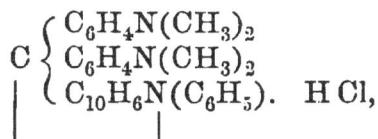

$$C \begin{cases} C_6H_4N(CH_3)_2 \\ C_6H_4N(CH_3)_2 \\ C_{10}H_6N(C_6H_5). \end{cases} \quad HCl,$$

according to which it is represented as the hydrochloride of tetramethyl-phenyl-triamido-diphenyl-naphthyl-carbinol hydro-chloride.

It forms a dark-blue powder with coppery reflex, which dissolves in water with a deep-blue colour. On boiling, the solution is rendered turbid and the free colour base is gradually thrown down as a reddish resinous precipitate. In presence of acetic acid this decomposition does not take place. Dilute sulphuric acid added to the aqueous solution turns it first green and then orange, but on neutralising the original blue colour is restored. This change is no doubt due to the formation of salts containing more than one equivalent of acid.

Victoria blue B may be dyed on wool or silk in a bath acidulated with acetic acid. The colours obtained are very fast to soap, but not to light. Cotton is first mordanted with Turkey-red oil and aluminium acetate.

Detection on the fibre.—Sulphuric acid changes the colour of the fibre to a reddish-brown, which is, however, restored to the original on washing with water.

Victoria blue 4 R is obtained by the action of methyl-

phenyl-α-naphthylamine on tetramethyldiamidobenzophenon and probably possesses the formula—

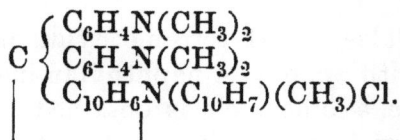

$$C \begin{cases} C_6H_4N(CH_3)_2 \\ C_6H_4N(CH_3)_2 \\ C_{10}H_6N(C_{10}H_7)(CH_3)Cl. \end{cases}$$

It dyes a much. redder shade than Victoria blue B,
but otherwise resembles it closely in application and properties.

Night blue.—This colouring matter is obtained by the
action of paratolyl-α-naphthylamine on tetramethyldiamidobenzophenon and is represented by the formula—

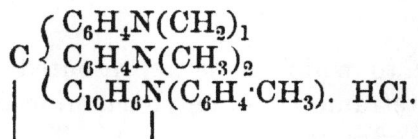

$$C \begin{cases} C_6H_4N(CH_2)_1 \\ C_6H_4N(CH_3)_2 \\ C_{10}H_6N(C_6H_4 \cdot CH_3). \end{cases} HCl.$$

Night blue forms a violet powder with a bronze reflex.
It is soluble in water with a blue colour and is completely precipitated from its aqueous or slightly acid
solution by picric acid, naphthol yellow and many other
acid colouring matters.

In its application and properties night blue resembles
Victoria blue B. It dyes an extremely pure shade of blue,
which is fast to soap on wool or silk.

Maroon (*Chestnut-brown*).

The resinous magenta residue is boiled with dilute
hydrochloric acid, as in the manufacture of phosphine.
After filtering, without further separation, all the bases
are precipitated with milk of lime. After washing, the
colour is sold as "maroon paste," or is neutralized with
hydrochloric acid, and thus rendered soluble in water.

The various sorts of maroon give various reactions, according to the proportion of the bases contained in them. The solutions of the hydrochloric acid compound in water may vary in colour from red to brown-violet. Hydrochloric acid colours it yellow to brown ; ammonia produces a dark, amorphous, flocculent precipitate, the solution being nearly colourless.

The blue-violet colouring matter contained in maroon, which is the hydrochloride of mauvaniline, may be prepared by dissolving in water or dilute hydrochloric acid, and isolating by fractional precipitation with salt.

The mauvaniline salt is precipitated first, and the rosaniline and chrysaniline remain in solution. The precipitate is purified by crystallising from boiling water.

The salts of mauvaniline crystallise well, and possess a beautiful bronze reflex. They dissolve sparingly in water, with a light blue-violet colour. They are easily soluble in hot water, especially after the addition of a little acid. Alkalies and ammonia precipitate the free base, which possesses the formula $C_{19}H_{17}N_3 \cdot H_2O$. Pure mauvaniline gives beautiful and fast violet shades, but it cannot be used along with other dye-stuffs.

Maroon gives a fine chestnut-brown. Silk is dyed in a bath containing boiled-off liquor, wool in pure water. Cotton is first mordanted with sumach or other tannin matter.

Aldehyde-green.

This colouring matter is prepared by mixing a solution of rosaniline in sulphuric acid with aldehyde, and warming, till a sample dissolves in water with a beautiful green colour. It is then poured into a boiling solution of hyposulphite of soda, boiled for some time, and filtered. Silk and wool may be dyed directly in the filtrate.

Aldehyde-green is now entirely replaced by other aniline greens.

(b) INDULINES AND SAFRANINES.

This division of basic dyes includes two small groups of dye-stuffs, which neither resemble each other in their relation to the fibres, nor in their chemical constitution. The only reason why they are described together is that the representatives of the safranines, as well as those of the indulines, may be obtained by the combination of the same mother substances.

If an amine, such as aniline, is allowed to act upon an amidoazo compound, as amidoazobenzene (aniline-yellow), so that a compound is formed with evolution of ammonia, an induline is formed :—

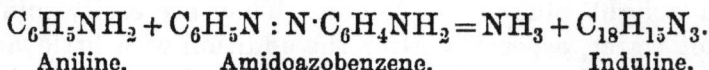

$$C_6H_5NH_2 + C_6H_5N : N \cdot C_6H_4NH_2 = NH_3 + C_{18}H_{15}N_3.$$

  Aniline.   Amidoazobenzene.   Induline.

If, however, in the place of ammonia, hydrogen is liberated,—in other words, if the mixture of amine and amidoazo body is oxidised,—safranine is formed:—

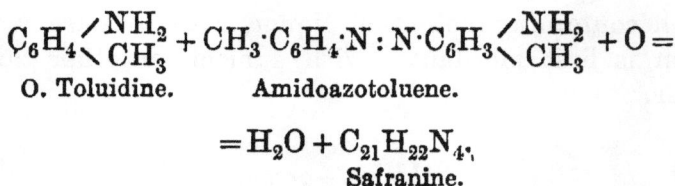

$$C_6H_4{<}^{NH_2}_{CH_3} + CH_3 \cdot C_6H_4 \cdot N : N \cdot C_6H_3{<}^{NH_2}_{CH_3} + O =$$

  O. Toluidine.   Amidoazotoluene.

$$= H_2O + C_{21}H_{22}N_4.$$

        Safranine.

INDULINES.

Induline, Nigrosine, Indigen, Coupier's Blue.

Violaniline.—In the blue, bluish-grey, grey-violet, or black-coloured indulines and nigrosines of commerce, there are a number of colouring matters which are seldom prepared in the pure state. They are, however, all closely related to one well-studied base—violaniline.

Violaniline is formed, as has been stated on one or two previous occasions, as a bye-product in the manufacture of magenta, by the arsenic-acid or nitrobenzene process.

In order to prepare it, magenta residues are boiled with hydrochloric acid, when the violaniline hydrochloride remains insoluble, along with resinous matter. In order to remove the latter, it is dissolved in hot aniline and filtered. On cooling, pure violaniline hydrochloride is precipitated. Violaniline is formed by the oxidation of aniline free from toluidine :—

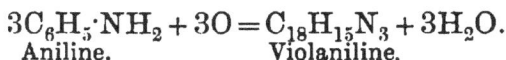

$$3C_6H_5{\cdot}NH_2 + 3O = C_{18}H_{15}N_3 + 3H_2O.$$
Aniline.   Violaniline.

It may also be obtained by a direct process—viz., by heating aniline and nitrobenzene with iron, a process greatly resembling the one used for magenta. The only differences are, that different temperature and proportions, and aniline free from toluidine, are used. If hydrochloride of aniline is heated, in alcoholic solution, with amidoazobenzene to 160°, a blue is obtained, according to the above equation, which is apparently identical with violaniline, and is known as *azodiphenyl blue*.

Properties of violaniline.—The hydrochloride—

$$C_{18}H_{15}N_3HCl,$$

is an amorphous, bluish-black powder, insoluble in water, soluble in alcohol. By adding caustic alkalies to the alcoholic solution, the free base is precipitated in flakes.

The alcoholic solutions of the spirit soluble nigrosines and indulines are used in the preparation of black spirit varnishes. They are also employed to some extent in calico-printing, and are fixed either by means of acetin or ethyltartaric acid.

Indulines and nigrosines soluble in water, Fast blue R, Fast-blue B, Blackley-blue, Guernsey-blue, Indigo substitute. —Violaniline contains, like rosaniline, hydrogen atoms which are replaceable by organic radicals. Spirit-soluble induline is obtained by phenylising violaniline. The solubility in alcohol of the product obtained stands in a direct ratio to the number of phenyl groups introduced. The highest substituted product is triphenylviolaniline, $C_{18}H_{12}(C_6H_5)_3N_3$. These colouring matters may be rendered soluble in water by treating them with sulphuric acid. The soluble indulines of commerce are generally mono- and disulphonic acids.

The manufacturer has it in his power, by varying the proportions, etc., in the nitrobenzene process, to prepare dye-stuffs of various shades and solubility.

The addition of certain metallic salts (zinc or copper chloride, etc.) has an influence on the shades of the product. The composition of the aniline oil alters the product, toluidine producing browner shades. The temperature to which the melt is heated in the process of manufacture is 230° C. By observing certain precautions the product may be obtained directly soluble in water; others are sulphonated afterwards.

Properties.—As shown above, the indulines and nigrosines vary considerably in their composition, and consequently their reactions are not in all cases identical. The following rules, however, hold good for both :—

Hydrochloric acid produces in the blue-violet or blue solution a similarly coloured precipitate; alkalies and ammonia redden the colour, but no precipitate is formed; tin-salts give a blue precipitate; zinc powder and ammonia decolourise the solution, especially on warming, but the original colour is rapidly restored on exposure to the air.

Application.—Silk is dyed in a bath containing boiled-off liquor, and is afterwards brightened with acetic acid.

Wool may be dyed in the same way as with alkali-blue —viz., first in an alkaline bath, and then developed with acid. A better method is to treat the wool first with chloride of lime solution, pass it into hydrochloric acid, and then dye in a bath containing sulphuric acid. Indulines and nigrosines yield very fast colours.

Detection on the fibre.—The colours resist the action of acids. Hydrochloric acid turns it, indeed, somewhat bluer, but nitric acid is almost entirely without action. Ammonia and caustic soda take up the colour, forming a red-violet solution, which is decolourised by zinc powder; the colour is restored on filtering and exposure to air Stannous chloride and hydrochloric acid strip the colour, forming a green solution. Some nigrosines are decolourised by chloride of lime, others are coloured reddish grey.

SAFRANINES.

According to the most recent investigations of Witt, Nietzki, and others, the safranines must be regarded as derivatives of the hypothetical base *azonium*, and the general formula for the chlorides would be represented thus:—

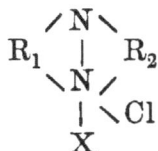

$$R_1 \underset{\displaystyle \underset{X}{\overset{\displaystyle |}{\underset{\displaystyle |\,\diagdown Cl}{}}}}{\overset{\diagup N \diagdown}{\underset{\diagdown N \diagup}{|}}} R_2$$

in which R_1 and R_2 each represent a bivalent and X a monovalent radical.

Phenosafranine or *Safranine B extra* is the simplest representative of this class of dyes, and is obtained by

the oxidation of a mixture of one molecule paraphenylenediamine and two molecules aniline. It is para-amidophenyl-
para-amidophenazonium chloride and its constitution is
represented thus :—

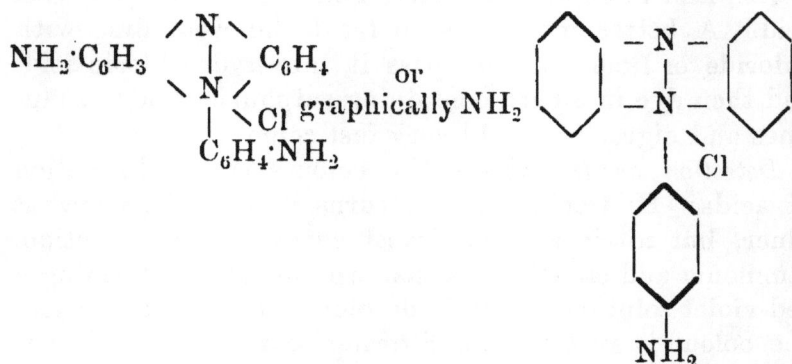

$$NH_2 \cdot C_6H_3 \begin{array}{c} \diagup N \diagdown \\ | \quad\quad C_6H_4 \\ \diagdown N \diagup \\ | \quad \diagdown Cl \\ C_6H_4 \cdot NH_2 \end{array} \quad \text{or} \quad \text{graphically}$$

The commercial product forms green glittering crystals
which dissolve in water with a red colour, and resemble
ordinary safranine in properties and application.

Safranine, Safranine T, Pink.

Safranine derives its name from the French safranon
(safflower). *Preparation.*—It was formerly obtained by
the oxidation of a mixture of amidoazotoluene and toluidine.

Amidoazotoluene is obtained by passing nitrous acid
into aniline oil rich in ortho-toluidine ("aniline for safranine").

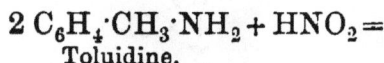

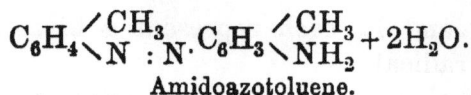

$$2 C_6H_4 \cdot CH_3 \cdot NH_2 + HNO_2 =$$
Toluidine.

$$C_6H_4 \diagup^{CH_3}_{\diagdown N} : N \cdot C_6H_3 \diagup^{CH_3}_{\diagdown NH_2} + 2H_2O.$$
Amidoazotoluene.

The product is mixed with the aniline oil, and carefully

oxidised with arsenic acid, then washed, and the oxidation completed by boiling with potassium bichromate. The boiling solution is mixed with milk of lime, which precipitates arsenic compounds, a violet bye-product, and chromic hydrate, while the safranine remains in solution. The whole is filtered, the filtrate neutralised with hydrochloric acid, and the dye-stuff salted out. The last impurities are removed by dissolving in water and salting out again.

According to a more recent and improved process, safranine is manufactured by the oxidation of a mixture of monamines and diamines. As in the old process, the aniline oil is converted into amidoazo compounds (amidoazobenzene and amidoazo-orthotoluene). The product is treated with zinc and hydrochloric acid, and there are formed, on the one hand, aniline and ortho-toluidine, and on the other, paraphenylenediamine, $C_6H_4(NH_2)_2$, and toluylene diamine, $C_6H_3 \cdot CH_3(NH_2)_2$. For amidoazotoluene the reaction takes place according to the equation—

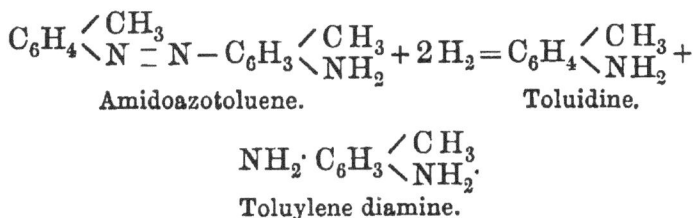

$$C_6H_4{<}^{CH_3}_{N\,=\,N}-C_6H_3{<}^{CH_3}_{NH_2}+2\,H_2 = C_6H_4{<}^{CH_3}_{NH_2}+$$

Amidoazotoluene.　　　　　　　　　　　Toluidine.

$$NH_2 \cdot C_6H_3{<}^{CH_3}_{NH_2}.$$

Toluylene diamine.

The product of the reaction is diluted with water, one molecule of hydrochloride of toluidine added, and the whole oxidised with potassium bichromate. The base is purified as in the old process.

Properties.—Safranine comes into commerce as a brown-red powder.

The pure salt forms fine, reddish crystals, with a green reflex, soluble in water and alcohol with a red colour. The alcoholic solution shows a beautiful yellowish fluorescence. It is insoluble in ether. Ammonia and alkalies produce no considerable change in colour, and no precipitate, the base being easily soluble in water.

To prepare the pure base, a solution of the hydrochloride is digested with silver oxide, the silver chloride filtered off, and the filtrate carefully evaporated to dryness. The residue acts very much like the hydrochloride. With picric acid, safranine forms a compound insoluble in water.

Concentrated sulphuric acid turns an aqueous solution of safranine blue, excess of strong acid turns it violet, and then green. Zinc powder and acetic acid reduce it even in the cold, leucosafranine being formed; but the solution becomes red again on exposure to the air.

Ordinary safranine consists of a mixture of tolusafranines and phenosafranines, $C_{21}H_{21}N_4Cl$, $C_{20}H_{19}N_4Cl$, and $C_{19}H_{17}N_4Cl$, of which the following may serve as a typical representative—

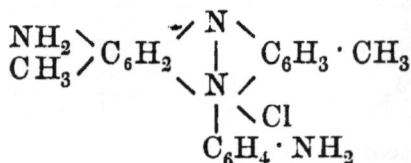

$$\begin{array}{c} NH_2 \\ \diagdown \\ CH_3 \diagup \end{array} C_6H_2 \begin{array}{c} \diagup N \diagdown \\ | \\ \diagdown N \diagup \\ | \diagdown Cl \\ C_6H_4 \cdot NH_2 \end{array} C_6H_3 \cdot CH_3$$

Application.—Safranine belongs to the small group of dye-stuffs which are taken up by animal fibres in alkaline solutions. Silk is dyed in a bath containing boiled-off liquor, wool in pure water.

In alkaline or neutral solutions, safranine possesses some affinity for vegetable fibres, but the colours produced are not fast. Cotton is therefore mordanted first. The best results are obtained by the use of tannin and tartar emetic.

Detection on the fibre.—Alcohol takes up the colour, forming a red solution which possesses a yellow fluorescence. Dilute hydrochloric acid is without action, while concentrated acid colours it blue-violet. Ammonia and caustic soda remove the colour, but without much alteration. Stannous chloride and hydrochloric acid decolourise on warming.

Naphthalene-red, Magdala-red, rosa-naphthylamine.

Preparation.—This beautiful dye-stuff is obtained from naphthylamine, $C_{10}H_7NH_2$, in the same manner as azodiphenyl-blue is derived from aniline. It is prepared as follows:—

Alpha-naphthylamine is first converted into amidoazonaphthalene, $C_{10}H_7N : N \cdot C_{10}H_6 \cdot NH_2$; for which purpose naphthylamine hydrochloride is dissolved in water, and mixed with a solution of caustic soda and sodium nitrite. The formation takes place according to the equation—

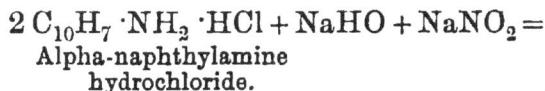

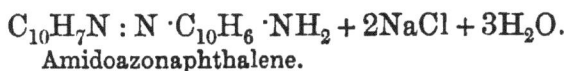

$$2\,C_{10}H_7 \cdot NH_2 \cdot HCl + NaHO + NaNO_2 =$$
Alpha-naphthylamine
hydrochloride.

$$C_{10}H_7N : N \cdot C_{10}H_6 \cdot NH_2 + 2NaCl + 3H_2O.$$
Amidoazonaphthalene.

The amidoazo compound separates out. In the pure state it forms beautiful red needles with a green reflex. The product is finely powdered, and mixed with the requisite quantity of alpha-naphthylamine, and so much acetic acid, that on heating to 150° complete solution is effected. The mass contains a violet colouring matter and excess of alpha-naphthylamine, and it has therefore to undergo a purification. The melt is boiled with a great excess of hydrochloric acid and filtered. The filtrate is neutralised with soda, and salt is added. Naphthalene-red being sparingly soluble in water, is precipitated, whilst the violet colour and naphthylamine remain in solution. The precipitate is purified by crystallisation from alcohol.

Properties.—Magdala-red is the chloride of diamidonaphthylnaphthazonium, and has the composition $C_{30}H_{21}N_4Cl$. It is a dark-brown crystalline powder, which may be obtained in large crystals, with a greenish reflex, by crystallization. It is only sparingly soluble in boiling water, but dissolves in alcohol with a cherry-red colour, the solution showing a cinnabar-red fluorescence.

This property is the best reaction for recognising naphthalene-red, as it is only resembled by some resorcin derivatives, such as diazoresorufin. The latter differ, however, from naphthalene-red in their behaviour towards caustic alkalies.

Ammonia and caustic soda produce no precipitate; with soda the colour becomes violet, and the fluorescence disappears.

Dilute acids are almost without action.

Concentrated sulphuric acid dissolves it with bluish-black colour.

Zinc powder and acetic acid decolourise; the colour returns on exposure to the air.

Application.—Naphthalene-red is used in silk-dyeing, especially for light shades. It gives a beautiful pink colour, with strong fluorescence, which is especially beautiful on velvet. Artificial light affects neither the colour nor the fluorescence. Darker shades are not brilliant, and may be produced better and cheaper than with other colouring matters, such as eosin, safranine, etc.

Silk may be dyed in a bath containing pure soap, or boiled-off liquor. One part of dye serves to produce a light rose colour on 1,000 parts of silk. The shades may be brightened with acids (tartaric, sulphuric).

The colouring matter is not suitable for wool, and indeed, its high price would prevent it ever being used in wood-dyeing.

Detection on the fibre.—It is faster than magenta, eosin, and safranine, and is neither affected by dilute acids nor alkalies. Its fluorescence is very characteristic. Alcohol does not dissolve it from the fibre, a characteristic difference from eosin, which is readily removed.

The following colouring matters also belong, according to their chemical constitution, to the group of the safranines :—

Giroflé.—This colouring matter is obtained by the action of nitrosodimethylaniline hydrochloride on a mixture of metaxylidine and paraxylidine hydrochlorides. It is xylyldimethylamidophenylxylylazonium chloride, and has the empirical formula $C_{24}H_{27}N_4Cl$. The commercial product forms a brown paste or a greyish-green powder, which dissolves in water with a magenta colour. It dyes a red-violet, and is used in calico printing for toning colours printed with alizarin violet.

Basle-blue is tolyldimethylamidophenotolylimidonaphthazonium chloride, $C_{32}H_{29}N_4Cl$, or—

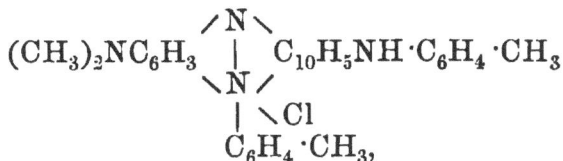

$$(CH_3)_2NC_6H_3 \underset{\diagdown N \diagup}{\overset{\diagup N \diagdown}{\underset{|\;\diagdown Cl}{|}}} C_{10}H_5NH\cdot C_6H_4\cdot CH_3$$
$$C_6H_4\cdot CH_3,$$

and is obtained by the action of nitrosodimethylaniline hydrochloride on ditolylnaphthylenediamine. It forms a brown crystalline powder, which dissolves in water with a blue-violet colour and dyes cotton mordanted with tannin and tartar emetic blue.

Neutral-blue is phenyldimethylpara-amidophenonaphthazonium chloride, $C_{24}H_{20}N_3Cl$, or—

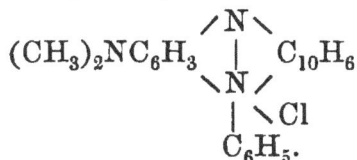

$$(CH_3)_2NC_6H_3 \underset{\diagdown N \diagup}{\overset{\diagup N \diagdown}{\underset{|\;\diagdown Cl}{|}}} C_{10}H_6$$
$$C_6H_5.$$

, Neutral-blue is formed by the action of nitrosodimethylaniline hydrochloride on phenylnaphthylamine. It

forms a brown amorphous powder, which dissolves easily
in water with a violet colour.

Mauveïn, Rosolan, Chrome-violet.

Violet colouring matters are obtained when aniline
containing toluidine is oxidised with chloride of lime,
permanganate of potash, lead peroxide, etc.

Amongst these is mauveïn, $C_{27}H_{25}N_4Cl$, the first aniline
dye introduced into commerce, which was discovered by
Perkin in 1856. He obtained it by the oxidation of aniline
oil containing toluidine with chromic acid. It is not much
used in dyeing at present.

The two following colouring matters belong to the
eurhodines, which are closely allied to the safranines in
their chemical constitution.

Neutral-violet is the hydrochloride of dimethyldiamido-
phenazine—

$$(CH_3)_2NC_6H_3 \underset{\diagdown N \diagup}{\overset{\diagup N \diagdown}{|}} C_6H_3 \cdot NH_2 \cdot HCl.$$

The commercial product forms a greenish-black powder, the dust of which, when inhaled violently, irritates the
mucous membranes, causing sneezing, etc. It dissolves
in water with a violet-red colour, and yields on cotton
mordanted with tannin and tartar emetic a red-violet.

Neutral-red is the hydrochloride of dimethyldiamido-
toluphenazine, $C_{15}H_{17}N_4Cl$, and is formed by the action of
nitrosodimethylaniline hydrochloride on metatolylene-
diamine. It dyes cotton mordanted with tannin and
tartar emetic bluish-red.

THE OXAZINES.

New-blue, Naphthylene-blue R, Fast-blue 2 B or R for

cotton, Cotton-blue R, Meldola's-blue, is obtained by the action of nitrosodimethylaniline hydrochloride on β naphthol. It is the chloride of dimethylphenylammonium-β-naphthoxazine, $C_{18}H_{15}N_2OCl$, or—

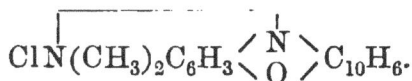

$$Cl\overset{\shortmid}{N}(CH_3)_2C_6H_3\Big\langle\overset{\shortmid}{\underset{O}{N}}\Big\rangle C_{10}H_6.$$

The commercial product forms a dark-violet powder with bronze reflex, the dust of which violently attacks the mucous membranes. It dissolves in water with a blue-violet colour, and yields on cotton mordanted with tannin and tartar emetic, or, better still, with tannin and iron, a deep indigo-blue shade.

Muscarin is the chloride of dimethylphenylpara-ammonium-β-oxynaphthoxazine, $C_{18}H_{15}N_2O_2Cl$, and is obtained by the action of nitrosodimethylaniline hydrochloride on a dioxynaphthalene. Its constitution is represented by the formula—

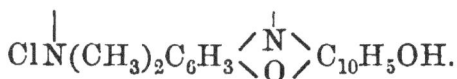

$$Cl\overset{\shortmid}{N}(CH_3)_2C_6H_3\Big\langle\overset{\shortmid}{\underset{O}{N}}\Big\rangle C_{10}H_5OH.$$

Muscarin forms a brown-violet powder, sparingly soluble in cold but easily soluble in hot water, with a blue-violet colour. The aqueous solution is decolourised by zinc powder, but the colour is restored on exposure to the air. It dyes cotton mordanted with tannin and tartar emetic blue.

Nile-blue is the sulphate of dimethylphenylpara-ammonium a amido-naphthoxazine, $(C_{18}H_{16}N_3O)_2SO_4$. The constitution of the corresponding hydrochloride is represented by the formula—

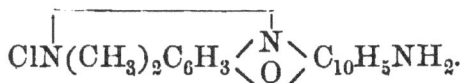

$$Cl\overset{\shortmid}{N}(CH_3)_2C_6H_3\Big\langle\overset{\shortmid}{\underset{O}{N}}\Big\rangle C_{10}H_5NH_2.$$

Nile-blue is formed by the action of nitrosodimethyl-meta-amidophenol on α naphthylamine.

The commercial product forms a dull-green crystalline powder, which dissolves easily in hot water with a blue colour. By the addition of hydrochloric acid to the aqueous solution, the hydrochloride is precipitated in minute needle-shaped crystals, which appear violet in transmitted light and green in reflected light. Caustic soda produces a red precipitate, which is taken up by ether, producing a brown-orange solution with a characteristic dark-green fluorescence.

Nile-blue dyes wool and silk in a neutral bath a red shade of blue. Cotton is previously mordanted with tannin and tartar emetic.

Gallocyanine, Solid-violet, also belongs to this class, and is the chloride of dimethylphenylammoniumdioxyphenoxazine carboxylic acid, $C_{15}H_{13}N_2O_5Cl$, or—

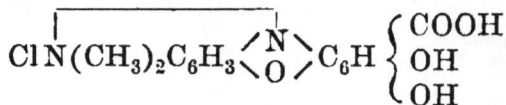

$$Cl\,N(CH_3)_2C_6H_3 \Big\langle {\overset{N}{\underset{O}{}}} \Big\rangle C_6H \begin{cases} COOH \\ OH \\ OH \end{cases}$$

Gallocyanine is formed by the action of nitrosodimethylaniline hydrochloride on gallic acid or tannic acid. It comes into commerce as a greenish paste, which is insoluble in water, but dissolves in alcohol with a blue-violet colour. In can be dyed directly on wool in presence of sulphuric acid, but much better results are obtained if the wool is previously mordanted with bichromate of potash. Gallocyanine is also used along with a chrome mordant in calico-printing. It produces blue shades of violet.

Prune is the methyl ether of gallocyanine.

INDOPHENOLS.

These colouring matters are produced by the simultaneous oxidation of a phenol and a paradiamine. This process resembles the manufacture of safranine, the only difference being that a phenol is oxidized with a para-diamine instead of a monamine with a paradiamine.

The following substances are used in the manufacture of indophenols:—

Phenol Paraphenylenediamine
Resorcin Dimethylparaphenylene-
Alpha- and beta-naphthol diamine.

Preparation.—Commercial indophenol is prepared as follows:—One molecule of nitrosodimethylaniline—

$$C_6H_4 \cdot NO \cdot N(CH_3)_2,$$

is reduced in aqueous solution to dimethylparapheny-lene diamine, $C_6H_4 \cdot NH_2 \cdot N(CH_3)_2$, filtered, and mixed with a solution of two molecules of alpha-naphthol dissolved in soda. Bichromate of potash is then added, and acetic acid is gradually run in until the liquid shows an acid reaction, when the colouring matter is precipitated.

Properties.—Indophenol comes into commerce as a blue paste or powder (indophenol N). The dried paste has a coppery lustre, and very much resembles Guatemala indigo in appearance. It sublimates in needles.

Indophenol is an oxidation product of dimethylpara-amidophenyl-a-oxy-a-naphthylamine. It possesses the empirical formula, $C_{18}H_{16}N_2O$, and the constitution—

$$N \begin{cases} C_6H_4N(CH_3)_2 \\ C_{10}H_6O \end{cases}$$

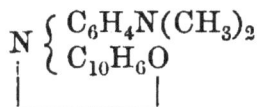

It dissolves in concentrated sulphuric acid, with a dingy yellow colour. It is insoluble in water, but soluble in alcohol, with a blue colour. Alkalies are without action on the solutions: acids colour them yellow.

Indophenol is reduced by glucose and caustic soda, forming a vat which resembles the indigo vat. This solution contains leucoindophenol, which is also a commercial article (*indophenol-white*, or indophenol preparation), and forms a white paste soluble in pure and in acidified water.

The reduced product has the composition—

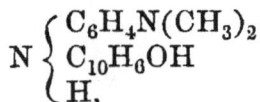

$$N \begin{cases} C_6H_4N(CH_3)_2 \\ C_{10}H_6OH \\ H, \end{cases}$$

and contains, therefore, two hydrogen atoms more than indophenol.

Application.—Indophenol gives very beautiful indigo-blue shades on cotton and wool. It is perfectly fast to light and weak chloride of lime solutions, but even weak acids decolourise it.

For printing on cotton, a mixture of indophenol and stannous hydrate (from stannous chloride and soda) is warmed with acetic acid until decolourised. It is then thickened with tragacanth, printed and steamed. The development of the leucoindophenol in the air is very slow, and it is therefore more advantageous to develop the colour in a bath of bichromate.

For dyeing wood and cotton, indophenol-white is dyed in neutral or slightly acid solution, and the colour subsequently developed with bichromate or chloride of lime.

When used along with indigo in the hyposulphite vat (with the addition of a little stannite of soda) it becomes fixed like indigo, and produces brighter and cheaper shades of blue than those obtained with indigo alone.

Detection on the fibre.—The reaction with dilute acids is most characteristic. A 10-per-cent. solution of hydrochloric acid turns it grey-brown or dark-grey, while other colouring matters are scarcely altered.

The colouring matters of this division are formed by the oxidation of amines of the aromatic series. They are characterised by their insolubility, and little tendency to crystallise. The dye-stuffs themselves are, on account of their insolubility, little suited for dyeing and printing. When fixed on the fibre, however, by means of oxidation, shades are obtained which are characterised by their extreme fastness.

Only two colouring matters of this division are technically applied—viz., aniline-black and the " direct naphthylamine-violet."

Aniline-black.

Preparation.—For the preparation of aniline-black in substance, aniline hydrochloride is very carefully oxidised. If the oxidation is carried too far, quinone, $C_6H_4O_2$, is produced; but if the oxidation is too feeble, green or violet colours are formed.

The following may be used as oxidising agents :—Chromic acid (potassium bichromate), permanganate of potash, and such metallic salts as easily give off oxygen. These metallic salts are generally employed along with potassium chlorate.

The most suitable combination is copper sulphate and potassium chlorate. The action of this mixture may be explained by the equation—

$$2KClO_3 + CuSO_4 = K_2SO_4 + Cu(ClO_3)_2.$$

The copper chlorate formed is very easily decomposed; its solutions give off gases at 60°, which consist essentially of chlorous anhydride.

A very useful material is a salt of vanadium, vanadiate of ammonia. In the formation of aniline-black, vanadium chloride is formed, which is immediately converted to vanadic acid by the potassium chlorate. One part of vanadium will do the work of 4,000 parts of copper, and will form from 10,000 to 200,000 parts of aniline-black.

Iron, chromium, and osmium salts may be used in place of copper sulphate, but do not possess any special advantages. Sulphate of cerium, however, is preferred by some to any other oxidising agent.

The finest aniline-black is obtained from pure aniline; ortho-toluidine gives a bluish-black, para-toluidine a brown-black. Aniline oil for black should boil at 182°. Aniline-black is also formed by the action of a galvanic current on a concentrated solution of an aniline salt. In this case, the oxidation is effected by the electrolytic oxygen evolved at the positive pole.

The following method serves for preparing aniline-black in substance:—Dissolve 40 parts of aniline hydrochloride, 20 parts of chlorate of potash, 40 parts of copper sulphate, and 16 parts of sal ammoniac, in 500 parts of water, and warm to 60°.

The crude aniline-black may be purified for analysis by boiling the precipitate obtained successively with hydrochloric acid, alcohol, ether, benzene, etc. The residue consists of the hydrochloride, from which the base is prepared by decomposition with dilute alkalies.

Properties.—Aniline-black has a formula consisting of some multiple of C_6H_5N, probably $C_{30}H_{25}N_5$.

The salts of aniline-black are unstable, as by drying or washing the acid is given off. To the hydrochloride the formula $C_{30}H_{25}N_5 \cdot 2\,HCl$ is usually given.[*]

In the dry state, aniline-black and its salts form black amorphous powders, which are insoluble in acids and alkalies. Crude aniline-black contains many other colour-

---

[*] According to Liechti and Suida, aniline-black possesses the formula $C_{18}H_{16}ClN_3$, in which the chlorine is not present as hydrochloric acid. It is thus represented as a chlorine-substitution product.

ing matters, which may be removed by different solvents. Thus boiling chloroform removes a blue-violet colouring matter, derived from ortho-toluidine, which possesses the empirical formula C_7H_7N.

Acetic acid, alcohol and benzene remove brown and red impurities.

Concentrated sulphuric acid dissolves aniline-black, with formation of a sulphonic acid. The new colouring matter is insoluble in acidulated water, and therefore separates out when the solution is poured into water. By continued washing with water, it dissolves with a green colour. This sulphonic acid serves for the preparation of an aniline-black vat, as its weak alkaline solutions are reduced by zinc-powder, grape-sugar, etc.

Fibres placed in this solution, and exposed to air, rapidly turn blue ; acids turn the colour green. By subsequent treatment with bichromate of potash, a fast grey is produced. Hitherto, this process has not found any technical application.

Strong oxidising agents, such as chromic acid, convert aniline-black into quinon. Energetic reducing agents completely destroy aniline-black, forming chiefly para-phenylenediamine, $C_6H_4(NH_2)_2$, and diamidodiphenyl-amine :—

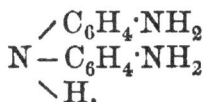

$$N \Big\langle \begin{matrix} C_6H_4 \cdot NH_2 \\ C_6H_4 \cdot NH_2 \\ H. \end{matrix}$$

Application.—Aniline-black comes into commerce as a paste, and finds a limited application as a steam black. The colour is thickened with albumen and fixed by steaming.

Aniline-black fixed direct on the fibre is of great importance in calico-printing, where it has to a great extent replaced the blacks formerly produced with logwood and madder.

In Lightfoot's original patent, granted in 1863, a thickened mixture of aniline, hydrochloric acid, copper chloride, and chlorate of potash is printed on the cotton. This method resembles the one given above for the preparation of aniline-black in substance, and produces a very good black.

The above mixture has, unfortunately, an injurious action on the steel "doctors" of the cylinder printing machine, because of the action of the copper salts on the iron, which is eaten away, while an equivalent quantity of copper is deposited.

Lauth overcame this drawback by using copper sulphide in place of the copper chloride. Lauth's method is used, with one or two variations, for ordinary aniline-black at the present time.

The colour consists of—

    10 litres starch paste.
    350 grams chlorate of potash.
    300   „   copper sulphate.
    300   „   sal ammoniac.
    800   „   aniline hydrochloride.

After printing, the fabrics are hung up in the aniline-black chamber, till a dark-green colour is developed. The temperature of this room is generally 30° to 40°, the degree of moisture being exactly regulated. The development generally lasts two days. The fabric is then passed through an alkaline bath of silicate of soda, chalk, or, if possible, ammonia. If the pieces are treated with ammonia, before the black is completely developed a blue shade is produced. Acids reproduce the green. This intermediate product is called emeraldine, and the blue produced by alkalies, azurine. Even the completely developed aniline-black appears to contain some emeraldine, as it becomes green when treated with acids. Aniline-black sometimes turn green if the pieces are allowed to lie in the air for some time. This alteration is most probably caused by sulphurous acid from the gas, where gas-lights are used. The black is reproduced, on washing, in its original beauty.

A black which does not turn green is obtained by a subsequent oxidation of the pieces which have been passed through ammonia. The best materials for this purpose are an acidified solution of bichromate of potash or a ferric salt. By this process iron or chromium lakes of aniline-black seem to be produced on the fibre, as the latter is found to absorb considerable quantities of the oxide unless great care is taken. Another method of obtaining aniline-blacks which do not turn green is based upon the fact that xylidine when used in place of aniline yields a black which after some time assumes a red cast. Pure aniline, as is well known, yields a black which turns green. If now the two bases are used together in the proper proportion, the green and the red which are subsequently formed neutralise each other, and the fabric consequently retains its original colour.

During the development of aniline-black, oxides of chlorine are formed, and these tend to injure the fibre. For very delicate fabrics, the hydrochloride of aniline may be replaced by a mixture of the tartrate with sal ammoniac.

In this process chlorate of soda is used instead of chlorate of potash, since the sparingly soluble potassium bitartrate would crystallise out, and either prevent or considerably retard the formation of black on the fibre.

The pieces printed by the above processes cannot be steamed, as the mixture would act violently on the fibre at such a temperature. Pieces have even been known to fire on steaming. Ordinary aniline-black cannot, therefore, be combined with steam colours. In producing mixed colours, the following process is adopted.

If it is desired to produce aniline-black and alizarin-red on a white ground, first print for the red with thickened acetate of alumina, then print the aniline-black, and develop in the usual manner. Pass through a chalk-bath, which effects the development of the black and the fixation of the alumina mordant. Then dye with alizarin and wash, clear, etc., in the usual way. (See Alizarin.) This process gives very beautiful results, but is rather difficult to manage.

A good steam black is obtained when aniline hydrochloride is completely replaced by aniline ferrocyanide. The chlorous acid developed in the nascent state acts on the hydroferrocyanic acid, producing hydroferricyanic acid.

On an aniline-black ground it is impossible to produce discharges, owing to the great stability of the colour. On this account, reserves have to be used. Reserves are generally alkaline, and act by the neutralisation of the acid colour, as aluminate or citrate of soda, etc.

Sulphocyanides form very suitable reserves; they absorb the chlorine, forming persulphocyanogen. The steam alizarin-red produced with aluminium sulphocyanide may be printed over with aniline-black, without being spoiled.

Aniline-black is as yet not suitable for wool- and silk-dyeing, since the feel, lustre and tenacity of the fibres are thereby injured.

Aniline-black is easily produced on cotton yarn. The dye-bath is prepared with potassium bichromate, aniline hydrochloride, and excess of hydrochloric acid, and the slightly soaped yarn is placed in the cold solution. After some time the bath is slowly heated to 60° C. The dyed goods are then washed in water, soda, and soft soap, to remove free acid, etc.

In this process it is very important to observe the proper proportions, and the concentration of the bath. If the right proportions are not adhered to, the aniline-black may be precipitated before the yarn is introduced, and a loss is thus caused. By using dilute solutions the black produced has a greyish cast.

The following proportions give good results :—

For 100 kg. yarn—
  Dissolve 2 kg. aniline in
        32 lit. hydrochloric acid, and
        42 lit. water,

pour into the dye-bath, containing a solution of 8 kg. potassium bichromate.

Detection on the fibre.—The colour is either unaltered by acids, or turned slightly greenish ; alkalies restore the original colour.

Alkalies are without action.

If passed several times through strong solutions of permanganate of potash and oxalic acid, alternately, the colour is destroyed. Chloride of lime turns the colour brown-red. Weak oxidising agents are without influence.

Naphthameïn.

(Naphthalene-violet.)

Alpha-naphthylamine, $C_{10}H_7NH_2$, yields by suitable oxidation naphthameïn, a violet colour, which is in many respects very similar to aniline-black.

It separates as an amorphous, purple precipitate, when an aqueous solution of hydrochloride of naphthylamine is treated with ferric chloride.

Naphthameïn is insoluble in water, dilute mineral acids, and alkalies, and is sparingly soluble in alcohol; but easily in acetic acid or ether. The acetic-acid solution may be used for dyeing and printing, but the colours produced with it are very dull.

Naphthameïn dissolves in concentrated sulphuric acid with a blue colour.

No formula has yet been assigned to this dye-stuff. A colouring matter similar, or identical, is obtained by printing with the aniline-black mixture, in which aniline is replaced by naphthylamine. The development and fixing is the same. The copper sulphide may be replaced by ferric chloride. After oxidation it is advantageous to pass the material through a solution of ferric chloride.

The shades produced vary from a grey-violet to a grey-brown. It is not nearly as fast to light, etc., as aniline-black.

Detection on the fibre.—Concentrated hydrochloric acid changes to a light grey; ammonia and caustic soda have little action.

(*d*) THIONINE COLOURING MATTERS.

Red, violet, and blue colouring matters containing sulphur are obtained when the hydrochlorides of some aromatic diamines are dissolved in sulphuretted hydrogen water and oxidised with ferric chloride.

The diamines used must have their amido groups in the para position to each other, ortho- and meta-diamido compounds yielding no colouring matters.

Of this class of colouring matters, only one—viz., methylene-blue—has become of technical importance, as it possesses a very beautiful colour, and the diamine, dimethylparaphenylenediamine, used in its preparation is more easily manufactured than the other diamines.

Methylene-blue.

Preparation.—The crude material for the manufacture of methylene-blue is dimethylaniline, $C_6H_5N(CH_3)_2$. This is dissolved in hydrochloric acid, and a solution of the exact equivalent of sodium nitrite added; para-nitroso-dimethylaniline is produced, according to the equation—

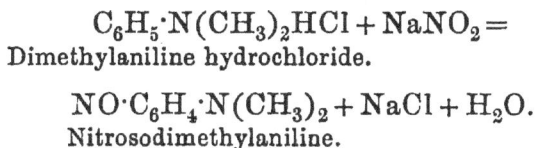

$$C_6H_5 \cdot N(CH_3)_2HCl + NaNO_2 =$$
Dimethylaniline hydrochloride.

$$NO \cdot C_6H_4 \cdot N(CH_3)_2 + NaCl + H_2O.$$
Nitrosodimethylaniline.

The liquid is then diluted with water and hydrochloric acid and placed in a vessel fitted with stirrer and flue, and is treated with sulphuretted hydrogen. The nitrosodimethylaniline is thereby reduced to dimethylparaphenylenediamine. When the reaction is over, the solution, which was yellow at first, is found to be completely decolourised.

The reaction is not a quantitative one, as by the action of the sulphuretted hydrogen a by-product is formed, which is converted to methylene-blue by mere exposure to air.

After decolourisation, small quantities of ferric chloride solution are added, till a slight excess of the latter is present. Then the solution is saturated with salt, and the dye separated by zinc chloride. A red colouring matter, not precipitated by zinc chloride, remains in solution. This colouring matter may be obtained as chief product, if the sulphuretted hydrogen is allowed to act for a longer period and more ferric chloride is used.

An interesting method of producing methylene-blue is described in a patent of Messrs. Ewer & Pick. It is based upon the following facts:—

If two plates of platinum, which constitute the two poles of an electric generator, are immersed in a solution of sulphuretted hydrogen, and paradimethylphenylenediamine

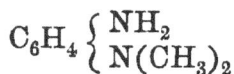

$$C_6H_4 \begin{cases} NH_2 \\ N(CH_3)_2 \end{cases}$$

in dilute sulphuric acid, an active development of hydrogen takes place at the negative pole; while the fluid which surrounds the positive pole becomes blue. The reaction ceases, however, in a short time, the positive pole becoming coated with a grey deposit. But if the plate is kept clean by means of a small brush, the whole of the sulphuretted hydrogen disappears, and the liquid begins to assume a blue colour. It is then found to contain chiefly reduced methylene-blue (methylene-white), along with a small quantity of methylene-blue in solution.

Properties.—Commercial methylene-blue is the chloride, or the zinc chloride, double salt of tetramethylthionin, $C_{16}H_{18}N_3S\cdot Cl$, or $2(C_{16}H_{18}N_3SCl) + ZnCl_2 + H_2O$.

The constitutional formula of the hydrochloride is represented thus :—

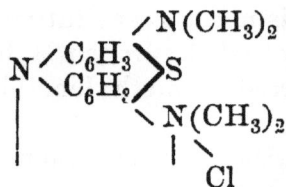

$$N<^{C_6H_3}_{C_6H_2}>S<^{N(CH_3)_2}_{N(CH_3)_2}$$
$$Cl$$

It forms a dark-blue or reddish-brown powder, easily soluble in alcohol or water. The *chloride*, $C_{16}H_{18}N_3\cdot S\cdot Cl$, may be prepared from the zinc compound by evaporating the aqueous solution to dryness, dissolving the residue in water, and adding concentrated hydrochloric acid. It forms dark-blue leaflets, easily soluble in alcohol and in water.

Ammonia does not precipitate methylene-blue solutions ; potash and soda produce blue precipitates.

The free base forms green needles with a metallic lustre, and may be crystallised from hot water.

Hydrochloric acid and nitric acid turn the solutions of methylene-blue greenish. Methylene-blue is easily decolourised by reducing agents, such as stannous chloride, zinc and acetic acid, etc. With tannin it forms a compound soluble in water, which is taken up by metallic mordants.

Application.—Methylene-blue has no special value for wool- and silk-dyeing, as for these materials other dye-stuffs are used which possess greater fastness and brilliancy.

On the other hand, it is of considerable importance in cotton-yarn dyeing, and still more so in calico-printing. The blue produced is of a greenish shade, and possesses great fastness. In artificial light the shades appear greenish.

Tannin and various mineral mordants are used in fixing ; the mordant depends on the shade required.

For pure blue, the following method may be used :— Mordant the oiled goods with alum, and fix the alumina in a chalk bath containing some sodium arseniate. Then heat in a weak tannin-bath, and dye the mordanted goods in a dye-bath containing methylene-blue, phosphate of soda and soda, beginning cold and raising the temperature gradually to the boil.

For dark-blue (indigo) :—The goods are mordanted after oiling in acetonitrate of iron. They are then placed in an ageing chamber, and the mordant is finally fixed in a chalk bath. On treating with tannin, a dark ground is produced, on which methylene-blue dyes a beautiful blue resembling indigo.

The tartar-emetic process is also applicable for methylene-blue. The methylene-blue and tannin, thickened with gum tragacanth, is printed and steamed, and subsequently fixed in a bath of tartar-emetic.

Detection on the fibre.—Methylene-blue is faster on cotton than aniline-blue. It resists the action of neutral soaps and dilute chloride of lime solutions, and is very fast light.

Ammonia is without action, but caustic alkalies and alkaline soaps remove the blue colour.

Hydrochloric acid takes up the blue, forming a green solution.

The following reactions are very characteristic :—Moistened with *hydrochloric acid*, the sample first turns green, and is gradually decolourised. Stannous chloride and other reducing agents reduce methylene-blue much quicker than other blue colouring matters. It is especially sensitive towards chromic acid. A three per cent. solution of potassium bichromate renders it violet, and finally discharges it. If fixed with tannin, a dark catechu-brown is produced.

Ethylene-blue is, according to Schulz and Julius (Tabell. Uebers. d. künstl. org. Farbstoffe), an impure methylene-blue, obtained by transforming nitrosodimethylaniline in strong sulphuric acid solution into the leucobase of methylene-blue by means of zinc sulphide and oxidising the product obtained.

Lauth's violet, or *Thionin*, is the hydrochloride of thionin—

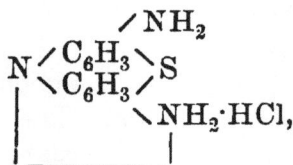

$$N \underset{\diagdown C_6H_3 \diagup}{\overset{\diagup C_6H_3 \diagdown}{\lessgtr}} S \underset{\diagdown NH_2 \cdot HCl,}{\overset{\diagup NH_2}{}}$$

and is obtained by the oxidation of paraphenylene diamine in presence of sulphuretted hydrogen with ferric chloride. It forms a dark-green powder with metallic reflex, which dissolves in water with a violet colour.

II. PHENOL DYE-STUFFS.

The phenol dyes are of an acid nature, which is due to the hydroxyl group or groups contained in them.

They are taken up by animal fibres, either in the free state, or in form of their soluble salts. Many give insoluble lakes with metallic oxides, and may therefore be dyed on animal fibres mordanted with alumina, iron, lead, etc. Other colouring matters of this group are applied with oil and tannin mordants.

The phenol dye-stuffs may be classified as follows :—

(*a*) Nitro-derivatives.

(*b*) Colouring matters produced by the action of nitrous acid on phenols.

(*c*) Rosolic acids.

(*d*) Phthaleïns.

In addition to the colouring matters described under section (*a*), a nitro-derivative of diphenylamine, *Aurantia*, will be considered.

This substance contains no hydroxyl group, and really belongs to the aniline dyes; but it also stands in close relation to the nitrophenols.

(*a*) NITRO-BODIES.

These dyes are as a rule of a yellow colour. They are stronger acids than the phenols from which they are derived.

They possess the following characteristics in common :— Strong acid reducing agents, such as stannous chloride and hydrochloric acid, convert them into the colourless amido derivatives. They dissolve in concentrated sulphuric acid, yielding either yellow or colourless solutions. Their solutions, or fibres dyed with them, are but slightly altered by hydrochloric acid, while ammonia and caustic soda tend to darken, or redden the colour. When dyed in acid baths, the colour may subsequently be partly removed from the fibre by boiling water. These reactions serve to distinguish these yellow colouring matters from others. Thus, phosphine is turned lighter by ammonia, while the yellow azo dyes are reddened by acids.

Picric Acid.

Preparation.—Picric acid or trinitrophenol may be obtained by warming pure carbolic acid with nitric acid :—

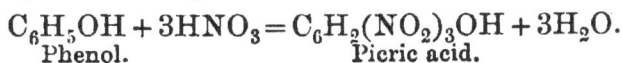

$$C_6H_5OH + 3HNO_3 = C_6H_2(NO_2)_3OH + 3H_2O.$$
Phenol.   Picric acid.

The process is much better, and less by-products are formed, if the phenol is first converted into its sulphonic acid. For this purpose, carbolic acid is heated with sulphuric acid to 100° till a sample dissolves completely in water. The liquid is then slightly diluted, run into strong nitric acid, and warmed. The reactions are expressed by the following equations :—

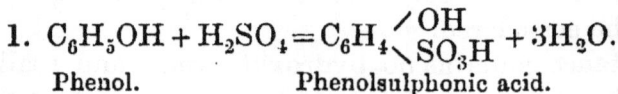

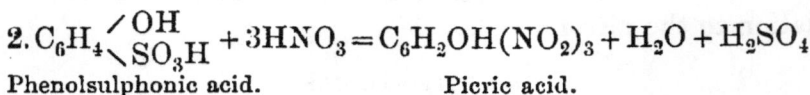

1. $C_6H_5OH + H_2SO_4 = C_6H_4\diagup{}^{OH}_{SO_3H} + 3H_2O.$
    Phenol.   Phenolsulphonic acid.

2. $C_6H_4\diagup{}^{OH}_{SO_3H} + 3HNO_3 = C_6H_2OH(NO_2)_3 + H_2O + H_2SO_4$
Phenolsulphonic acid.   Picric acid.

The acid liquor contains picric acid, nitric acid, sulphuric acid, and resinous impurities. It is diluted with water, and soda added till the resins are separated, after which it is filtered and excess of soda solution added. The precipitated sodium picrate is dissolved in water and decomposed with sulphuric acid. The picric acid is salted out.

Picric acid has the constitution :—

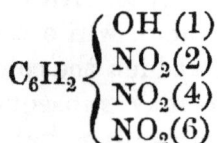

$$C_6H_2\begin{cases} OH\ (1) \\ NO_2(2) \\ NO_2(4) \\ NO_2(6) \end{cases}$$

Properties.—Picric acid forms light-yellow leaflets, m.p. 122·5°. It may be sublimed by cautious heating; rapid heating causes it to explode. It possesses an extremely bitter taste (hence its name, from πικρός = bitter). One part of picric acid dissolves in 86 parts of water at 15° and in 26 parts at 76°; it is easily soluble in alcohol, ether and benzene. The solutions are yellower than the free acid; the addition of a little sulphuric acid causes the colour to become much lighter.

Picric acid is a strong, monobasic acid; its salts are of a yellow or orange colour. They explode with great violence on heating. The potassium salt, $C_6H_2 \cdot (NO_2)_3OK$, is distinguished by its comparative insolubility, 1 part requiring 288 parts of water at 15° to dissolve it. The ammonium and sodium salts are more easily soluble. The salts with the alkaline earths are soluble in water. One part of the lead-salt dissolves in 119 parts of water at 15°.

By warming with tin and hydrochloric acid, picric acid is completely reduced; the colourless solution contains the hydrochloric acid compound of triamidophenol, $C_6H_2(NH_2)_3OH$.

If alkaline reducing agents are used, the reduction is not carried so far. Thus, if sulphuretted hydrogen is passed into a solution of picric acid in alcoholic ammonia, only one nitro group is reduced, and dinitroamidophenol, or picramic acid is produced:—

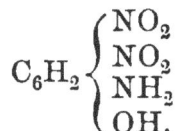

$$C_6H_2 \begin{cases} NO_2 \\ NO_2 \\ NH_2 \\ OH. \end{cases}$$

The latter dyes wool and silk brown.

If concentrated solutions of picric acid and potassium cyanide are mixed, a dark-red solution is formed, and on standing reddish-brown crystals are deposited. They consist of the potassium salt of isopurpuric acid, $C_8H_4KN_5O_6$, and are produced according to the equation—

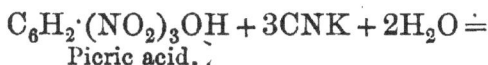

$$C_6H_2 \cdot (NO_2)_3OH + 3CNK + 2H_2O =$$
Picric acid.

$$C_8H_4KN_5O_6 + K_2CO_3 + NH_3.$$
Potassium isopurpurate.

Free isopurpuric acid is very unstable, and is only known in the form of salts. Its potassium or ammonium

salt was formerly used as a dye, under the name of
" Grenat soluble." It dyes a brown upon wool or silk.
The baths must not be strongly acid ; the addition of a
little acetic or tartaric acid is best.

Testing of picric acid.—Commercial picric acid is generally crystallised. It should dissolve in water, acidified
with sulphuric acid, without any residue. It should be
entirely soluble in 10 parts of alcohol. If alcohol leaves
a residue, the latter will most probably consist of inorganic
salts (Glauber's salts, nitre, etc.), which may be recognised
in the usual manner.

Oxalic acid is sometimes added. To detect it, the
sample is dissolved in ammonia, and calcium chloride is
added. A white precipitate indicates oxalic acid.

Sugar is detected by saturating the aqueous solution
with sodium carbonate, evaporating to dryness, and extracting with diluted alcohol ; sugar is dissolved while
picrate of soda remains. Pure picric acid should be easily
soluble in benzene.

The quantitative estimation of picric acid may be
effected by the method described on p. 65.

Application of picric acid.—Picric acid is a substantive
colour on silk and wool. Its dyeing power is very great,
one part being sufficient to dye a thousand parts of silk a
distinct yellow.

Animal fibres are dyed in acid baths. Sulphuric acid
is the most advantageous for this purpose, as it causes the
colour to deposit more evenly.

Picric acid yellow is darkened by the action of light
and air. It may be removed from the fibre by repeated
washing.

In order to fix it upon wool, alum or bichromate of
potash are sometimes used. The picrates of alumina and
chromium are soluble in a large quantity of water ; thus
no lakes are formed on the fibre, and the only advantage
of this method is that picric acid possesses a somewhat

greater affinity to alumina and chromium salts than to the free fibre.

A method of producing weighted yellow silks is based upon the fact that picric acid forms a sparingly soluble lead-salt. For this purpose the silk is first mordanted with lead acetate, and then dyed in picric acid. Silk containing picrate of lead is, however, very liable to take fire, the flame being very difficult to extinguish. The sulphuretted hydrogen in the air is also liable to blacken it, owing to the formation of lead sulphide.

On cotton, picric acid may be fixed with albumen; but the colours are so dull as to have no practical value.

Picric acid gives a somewhat greenish yellow, and thus is seldom used for pure yellow.

On the other hand, it is very suitable for dyeing mixed colours. It may be combined with methyl-green for yellow-green, indigo-carmine or aniline-blue for green, with violet for olive, and with magenta for scarlet.

A very good green is obtained on wool with indigo-carmine and picric acid.

The picrates of the rosaniline bases are sparingly soluble or insoluble.

Besides, in dyeing, picric acid is used extensively in the manufacture of explosives.

Detection on the fibre.—Picric acid yellow is reddened by a mixture of stannous chloride and alkali (formation of picramic acid), and also by potassium cyanide (formation of isopurpuric acid).

In order to detect it in fabrics dyed with mixed colours, the best method is to extract with dilute alcohol, evaporate to dryness, and test the residue with ammonium sulphide or potassium cyanide.

All fabrics dyed with picric acid possess a bitter taste.

In some cases picric acid is added to beer, to give it the required bitter taste. In order to detect an adulteration of this kind, 10 c.c. of the beer are shaken with

5 c.c. of amyl alcohol, the extract evaporated, and tested
with potassium cyanide or ammonium sulphide.

Phenicienne.   Phenyl-brown.

If phenol is nitrated by another process, instead of picric
acid, a brown colour, phenyl-brown, is obtained.

In order to prepare it, 10 to 12 parts of a mixture of
two volumes of sulphuric acid and one volume of nitric
acid, sp. gr. 1·35, are gradually added to 1 part of cooled
phenol.   A considerable amount of nitrous fumes is
evolved.   The product is then poured into water, collected
on a filter, and the precipitate washed.

Phenyl-brown consists of two substances.   The brown
portion is amorphous, insoluble in water, but soluble in
alkalies and spirits of wine.   Its chemical composition
is unknown.   The second constituent is a dinitrophenol,
$C_6H_3(OH)\cdot(NO_2)_2$; it dyes yellow, is crystallisable, and
dissolves in hot water, alkalies and alcohol.

Phenyl-brown is a yellowish-brown powder, which melts
on gentle heating, and on stronger heating deflagrates,
owing to the dinitrophenol it contains.   It is only partially insoluble in water, but dissolves completely in
alcohol.

The yellowish-brown solution is turned yellow by
hydrochloric acid, and after some time a precipitate is
produced.

Caustic soda and ammonia turn the liquid bluish violet.
Metallic salts produce precipitates.

Phenyl-brown is often used in dyeing leather.

On wool it produces Havanna-brown shades, which are
very fast to light.   The colours are spoiled by steaming.

By chroming, i.e., passing through a bath of bichromate
acidified with sulphuric acid, the colour is turned to a
ruby-red.

Victoria-yellow.

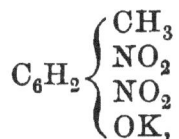

Victoria-yellow—

$$C_6H_2 \begin{cases} CH_3 \\ NO_2 \\ NO_2 \\ OK, \end{cases}$$

is the potassium salt of dinitroparacresol, and is produced by the action of nitric acid on paratoluidine, $C_6H_4 \cdot CH_3 \cdot NH_2$, or on paracresol, $C_6H_4 \cdot CH_3 \cdot OH$.

It forms red crystals, which dissolve in water with a yellow colour. Hydrochloric acid decolourises the solution, and precipitates the free acid in the form of light-yellow needles. Caustic alkalies and ammonia produce no alteration. By warming with potassium cyanide, a red colour is produced, similar in composition to isopurpuric acid.

Victoria-yellow produces yellow shades on wool and silk, which are somewhat redder than those obtained with picric acid. The colours are, however, so unstable that they are seldom used in dyeing.

Detection.—Warm water removes the colour from the fibre. The dilute yellow solution is decolourised by hydrochloric acid; if concentrated, a precipitate is formed.

Hydrochloric acid decolourises the fibre; water reproduces the original colour.

Naphthol-yellow.

(Martius-yellow. Manchester-yellow.)

Naphthol-yellow is the soda, potash, or lime-salt of dinitronaphthol :—

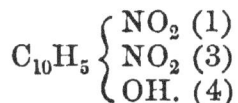

$$C_{10}H_5 \begin{cases} NO_2 \ (1) \\ NO_2 \ (3) \\ OH. \ (4) \end{cases}$$

Preparation.—Alpha-naphthol is treated at 100° with a mixture of nitric acid and sulphuric acid, and the nitro-compound formed is precipitated by pouring the product into water.

Or, the alpha-naphthol is converted into the monosulphonic acid with sulphuric acid, and then nitrated.

The sulpho group is thus removed, and substituted by NO_2, as in the preparation of picric acid.

Martius-yellow is also formed by the action of nitric acid on α naphthylamine.

Properties.—Dinitronapthol forms yellow needles, which are insoluble in water. It melts at 138°. It is a strong acid, forming yellow or orange salts.

The sodium salt, $C_{10}H_5(NO_2)_2ONa + H_2O$, forms needles which are easily soluble in water. The lime-salt possesses the formula $(C_{10}H_5(NO_2)_2O)_2Ca + 6H_2O$.

The solutions of naphthol-yellow are decolourised by hydrochloric acid, a precipitate of the free acid being produced. Ammonia is without action; caustic potash or soda produce precipitates of an orange-red colour. Potassium cyanide gives the isopurpuric-acid reaction; ammonium sulphide colours the solution red.

Naphthol-yellow is sometimes adulterated with picric acid, to detect which a sample is dissolved in water, and the dinitronaphthol precipitated by hydrochloric acid. In presence of picric acid the filtrate has a yellow colour. The picric acid may be crystallised on evaporation.

Application.—Naphthol-yellow gives a very beautiful golden-yellow on silk and wool, and is sometimes used for mixed colours. A great drawback to its application is its sensibility to heat. A very slight increase of temperature causes it to volatilise; this even takes place in a summer heat.

Silk and wool are dyed in a bath containing acetic acid.

Naphthol-yellow, like picric acid and Victoria-yellow, is not applicable to cotton.

Detection on the fibre.—Water takes up the colouring matter. The yellow solution is decolourised by dilute sulphuric acid (picric acid is only turned lighter). Boiling potassium cyanide gives it a red colour. Hydrochloric acid completely decolourises. If a sample of the material is wrapped in a piece of white paper and heated to 120° in an air-bath, part of the yellow colour is transferred to the paper.

Naphthol-yellow S.

Preparation.—Naphthol-yellow S is a sulphonic acid of Martius-yellow. The free acid has the formula—

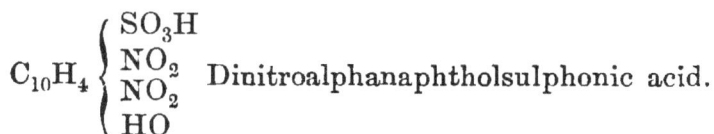

$$C_{10}H_4 \begin{cases} SO_3H \\ NO_2 \\ NO_2 \\ HO \end{cases} \text{Dinitroalphanaphtholsulphonic acid.}$$

By nitration of alpha-naphthol monosulphonic acid—

$$C_{10}H_6(SO_3H)OH,$$

the sulpho group is eliminated, and Martius-yellow is formed. If alpha-naphthol trisulphonic acid, $C_{10}H_4(SO_3H)_3OH$, is nitrated, two sulpho groups are eliminated, and naphthol-yellow S is formed.

Alpha-naphthol is warmed with twice its weight of sulphuric acid (containing 25 per cent. anhydride) to 40–50°, and the monosulphonic acid is then converted into the trisulphonic acid by the addition of sulphuric acid containing 70 per cent. of anhydride. The mass is diluted with a little water, and treated with concentrated nitric acid. On cooling, dinitronaphthol sulphonic acid crystallises out. It is purified by recrystallisation, and is converted into the ammonium or sodium salt. From the mother-liquor the sulphuric acid is removed by lime, and the rest of the colouring matter is precipitated by potash.

Properties.—The free acid forms long yellow needles easily soluble in hot water. It is a strong acid, sometimes replacing even sulphuric acid in its compounds. If, for instance, its aqueous solution is mixed with potassium sulphate, the sparingly soluble potassium salt is precipitated, while free sulphuric acid remains in solution.

The potassium salt—

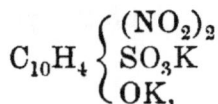

$$C_{10}H_4 \begin{cases} (NO_2)_2 \\ SO_3K \\ OK, \end{cases}$$

is sparingly soluble in cold water, more readily in hot water. If treated with strong sulphuric acid, the free acid is not precipitated, but an acid salt is formed, which possesses the formula—

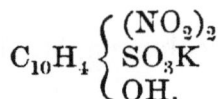

$$C_{10}H_4 \begin{cases} (NO_2)_2 \\ SO_3K \\ OH. \end{cases}$$

Application.—Naphthol-yellow S is faster to washing than picric acid and Martius-yellow. It does not volatilise on steaming. Silk and wool are dyed in a bath acidified with sulphuric acid. The addition of stannic chloride brightens the colour on wool.

Detection on the fibre.—Boiling water does not remove the colour; a sample does not stain at 110° to 120°. Otherwise the reactions are the same as those of naphthol-yellow.

Brilliant yellow is the sodium salt of a dinitro α naphtholmonosulphuric acid $C_{10}H_4OH(NO_2)_2SO_3Na$ obtained by the action of nitric acid on the alpha-naphtholtrisulphonic acid of the Schöllkopf Aniline Co. It resembles naphthol-yellow S.

Heliochrysin (Sun-gold).

This beautiful yellow colouring matter is the sodium salt of tetranitronaphthol :—

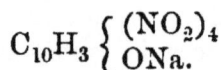

$$C_{10}H_3 \begin{cases} (NO_2)_4 \\ ONa. \end{cases}$$

It is obtained by energetic nitration of monobromnaph-thalene, $C_{10}H_7Br$, and warming the tetranitrobromnaph-thalene with soda solution :—

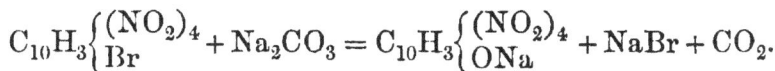

$$C_{10}H_3\begin{Bmatrix}(NO_2)_4\\Br\end{Bmatrix} + Na_2CO_3 = C_{10}H_3\begin{Bmatrix}(NO_2)_4\\ONa\end{Bmatrix} + NaBr + CO_2.$$

Tetranitrobromnaphthalene   Heliochrysin.

This colouring matter is not fast enough to light to be of much technical importance.

Citronine is a mixture of tetranitrodiphenylamine and dinitrodiphenylamine obtained by the action of nitric acid on diphenylamine. It produces a golden yellow on wool and silk in an acid bath.

Aurantia.

Diphenylamine, $N(C_6H_5)_2H$, or better, methyldiphenyl-amine, $N(C_6H_5)_2 \cdot CH_3$, yields, on warming with nitric acid, a yellow substance, insoluble in water—hexanitro-diphenylamine :—

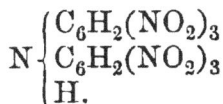

$$N\begin{Bmatrix}C_6H_2(NO_2)_3\\C_6H_2(NO_2)_3\\H.\end{Bmatrix}$$

This nitro product is a strong acid, the hydrogen atom bound to the nitrogen being easily replaced by metals.

The ammonium salt, $N(C_6H_2(NO_2)_3)_2 \cdot NH_4$, comes into commerce as aurantia.

Aurantia is a crystalline, reddish-yellow powder, which deflagrates slightly on heating. It is easily soluble in water and alcohol. The aqueous orange-yellow solution is darkened and reddened by alkalies. Acids precipitate the free acid as a sulphur-yellow, flocculent precipitate, the solution being rendered nearly colourless.

An acid solution of stannous chloride gives the same reaction ; but on boiling, the yellow precipitate becomes dark brown.   Copper salts also turn the solution browner.

Application.—Silk and wool are dyed in baths containing a little sulphuric acid.   The above remarks on the action of metallic salts render it necessary that aurantia must be dyed in vessels of wood or glass,

The application of aurantia has one very great drawback.   It is not entirely harmless.   Professor Gnehm, the discoverer, states, that even dilute solutions of aurantia produce very painful blisters on the skin.   The medical faculty of Cologne have, on the other hand, sanctioned its manufacture, as being harmless.

Detection on the fibre.—Hydrochloric acid turns it lighter yellow ; ammonia and alkalies produce little alteration. The most characteristic reaction is shown on warming with stannous chloride, which turns it dark-brown.

(b) COLOURING MATTERS PRODUCED BY THE ACTION OF NITROUS ACID ON PHENOLS.

Some phenols, such as carbolic acid, thymol, resorcin, orcin, and napbthol, are converted into dye-stuffs when treated with nitrous acid, in ethereal or sulphuric acid solution.

For this purpose the phenol is dissolved in strong sulphuric acid, and a solution of nitrous acid in strong sulphuric acid is added.   The nitrous solution is prepared by gradual addition of 5 per cent. of sodium nitrite to concentrated sulphuric acid.

The mixture is warmed on the water-bath till the reaction is finished, and the product is precipitated by pouring into water.

Phenol, $C_6H_5 \cdot OH$, treated as above, gives a brown flocculent precipitate, which dissolves in sulphuric acid and alkalies, with a beautiful blue colour.   It is known as " Liebermann's Phenol Dye-stuff."   From orcin colouring matters are obtained which are closely related to the orchil colours, especially orceïn.

Of all these colouring matters, the one derived from resorcin alone has technical importance.

Fluorescent Resorcin-blue.

Resorcin gives, when treated as above, a red substance, diazoresorufin, $C_{18}H_{10}N_2O_5$.

Diazoresorufin is also produced by warming mononitrosoresorcin with resorcin and sulphuric acid :—

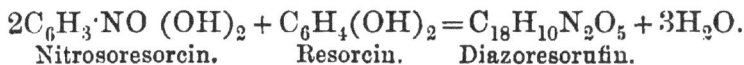

$$2C_6H_3 \cdot NO \, (OH)_2 + C_6H_4(OH)_2 = C_{18}H_{10}N_2O_5 + 3H_2O.$$
Nitrosoresorcin.   Resorcin.   Diazoresorufin.

This confirms the above formula.

Diazoresorufin is nearly insoluble in water and alcohol. It is a very weak acid, and forms salts with alkalies which dissolve in water with a cherry-red colour, the solution showing a beautiful cinnabar-red fluorescence.

Diazoresorufin itself is not suitable for dyeing, for the beautiful colours of the alkaline salts are only shown in neutral or alkaline solutions, while in acid baths wool and silk only assume a dirty-brown colour.

By the action of bromine on diazoresorufin, a hexabromdiazoresorufin is obtained, $C_{18}H_4Br_6N_2O_5$ (?), which resists the action of acids. Its ammonium-salt is the commercial resorcin-blue, or " bleu fluorescent."

It forms a 10 per-cent. paste, in which the green metallic needles of the colouring matter may be distinguished. It is sparingly soluble in water or absolute alcohol ; the best solvent is a mixture of equal parts of alcohol and water. The solutions are blue by transmitted light, red by reflected light. Strong acids, like hydrochloric or sulphuric acid, precipitate brown hexabrom-resorufin.

Silk is dyed with resorcin-blue in a neutral soap-bath. The paste is added to the bath, and dissolves easily.

Q

Brightening is effected by tartaric or sulphuric acid.

The colour produced upon silk and wool is blue with a slight admixture of red and grey, and has a characteristic red fluorescence, especially in artificial light. It is perfectly fast to light, washing, and acid.

By combination with a yellow colouring matter, a beautiful fluorescent olive is produced.

On the fibre, resorcin-blue is easily distinguished by its fluorescence, and the action of dilute sulphuric acid. Ammonia and alkalies take up the colour, forming a blue solution with red fluorescence.

Resorcin-green.　Dark-green.

This colouring matter is dinitrosoresorcin,—

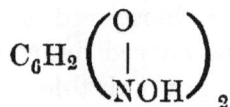

$$C_6H_2\left(\begin{array}{c}O\\|\\NOH\end{array}\right)_2$$

and is obtained by the action of nitrous acid on resorcin. The commercial product forms a dirty-green paste, which is sparingly soluble in water.

Application.—Cotton is mordanted with tannin and nitrate of iron; wool with copperas and tartar. Various shades of dark-green may thus be obtained which are characterised by their fastness.

Gambine.

Gambine is nitroso β naphthol. The commercial product forms a dirty-yellow or brown-yellow paste.

Application.—On wool mordanted with iron, gambine produces full shades of olive-green. With bichromate of potash as mordant, dull shades of yellow are obtained.

Naphthol-green.

This colouring matter is the ferrous-sodium salt of nitroso β naphtholmonosulphonic acid, and has the composition, $Fe(C_{10}H_5 \cdot ONO \cdot SO_3Na)_3$. It is prepared by first treating β naphtholsulphonic acid S with nitrous acid and combining the product found with soda and ferrous oxide.

Naphthol-green forms a dark-green powder which dissolves in water with a green colour.

Application.—Naphthol-green may be dyed on wool in presence of copperas and sulphuric acid. It produces full shades of olive-green which are quite fast to light.

(c) ROSOLIC ACIDS.

The rosolic acids are derivatives of triphenylmethane, and stand in close relation to the rosanilines. They are rosanilines, in which the amido groups are substituted by hydroxyl. A product corresponding to pararosaniline would thus be constituted as follows:—

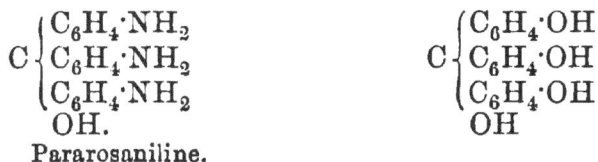

$$C \begin{cases} C_6H_4 \cdot NH_2 \\ C_6H_4 \cdot NH_2 \\ C_6H_4 \cdot NH_2 \\ OH. \end{cases} \qquad C \begin{cases} C_6H_4 \cdot OH \\ C_6H_4 \cdot OH \\ C_6H_4 \cdot OH \\ OH \end{cases}$$

Pararosaniline.

The intermediate product expressed by the last formula is, however, not known; it splits off water, forming pararosolic acid or aurin—

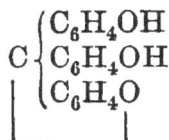

$$C \begin{cases} C_6H_4OH \\ C_6H_4OH \\ C_6H_4O \end{cases}$$

just like pararosaniline splits off water in uniting with acids.

Pararosaniline may be converted into aurin in the following manner.

The hydrochloride of pararosaniline treated in dilute solution with sodium nitrite, yields diazopararosaniline chloride :—

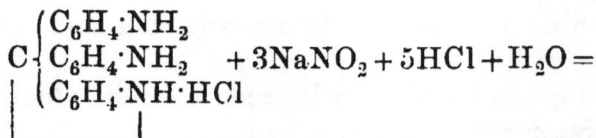

$$C \begin{cases} C_6H_4 \cdot NH_2 \\ C_6H_4 \cdot NH_2 \\ C_6H_4 \cdot NH \cdot HCl \end{cases} + 3NaNO_2 + 5HCl + H_2O =$$

Pararosaniline hydrochloride.

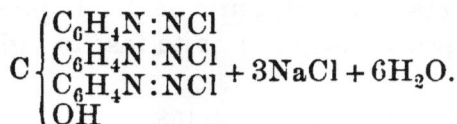

$$C \begin{cases} C_6H_4N : NCl \\ C_6H_4N : NCl \\ C_6H_4N : NCl \\ OH \end{cases} + 3NaCl + 6H_2O.$$

Diazopararosaniline chloride.

The liquid is then boiled, with the addition of sulphuric acid, when aurin is formed :—

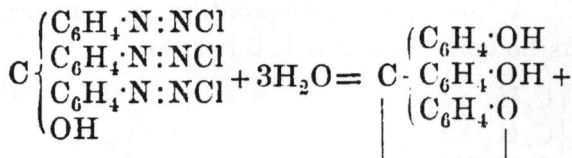

$$C \begin{cases} C_6H_4 \cdot N : NCl \\ C_6H_4 \cdot N : NCl \\ C_6H_4 \cdot N : NCl \\ OH \end{cases} + 3H_2O = C \begin{cases} C_6H_4 \cdot OH \\ C_6H_4 \cdot OH \\ C_6H_4 \cdot O \end{cases} +$$

Diazopararosaniline chloride.　　　　Aurin.

$$3N_2 + 3HCl + H_2O.$$

On the other hand, aurin heated to 200° with aqueous ammonia is reconverted into pararosaniline :—

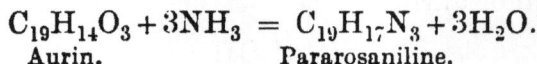

$$C_{19}H_{14}O_3 + 3NH_3 = C_{19}H_{17}N_3 + 3H_2O.$$
　　Aurin.　　　　　　Pararosaniline.

From rosaniline, rosolic acid may be prepared in a similar manner.

These two colouring matters are now only of secondary importance, but were formerly manufactured on a very large scale.

Corallin.

Yellow corallin.—To prepare this colouring matter, 8 parts of pure phenol are mixed in the cold with 3·2 parts of concentrated sulphuric acid, and after some hours 4·8 parts of oxalic acid are added, and the whole heated to 110° for twenty-four hours. The mass is then poured into water, and extracted several times with boiling water.

This product is yellow corallin. It contains about 20 per cent. of aurin, formed by the action of the nascent formic acid from the oxalic acid, upon phenol.

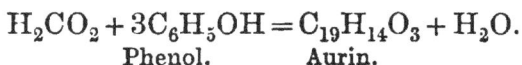

$$H_2CO_2 + 3C_6H_5OH = C_{19}H_{14}O_3 + H_2O.$$
$$\text{Phenol.} \qquad \text{Aurin.}$$

The other constituents are crystalline derivatives of rosolic acid, and resinous bodies, which are mostly colourless.

Yellow corallin is a brown mass possessing a green metallic lustre. It is almost insoluble in water, but easily soluble in alcohol. The yellow, alcoholic solution is turned red by alkalies, and yellow again by acids. This reaction is so delicate as to render corallin useful as an indicator in volumetric analysis.

Pure aurin, $C_{19}H_{11}O_3$, forms ruby-red crystals, with a blue fluorescence. It forms unstable salts with bases.

Red corallin.—If 2 parts of yellow corallin are heated with 1 part of ammonia liquor in a closed vessel to 120°–140°, a red colouring matter is formed, which is precipitated by pouring the product into water and acidifying.

It has been stated before, that at 200° pararosaniline is formed. Red corallin is an intermediate product, in which only one hydroxyl group of the aurin is replaced by NH_2:—

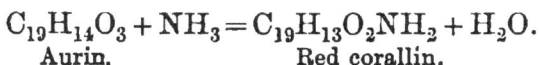

$$C_{19}H_{14}O_3 + NH_3 = C_{19}H_{13}O_2NH_2 + H_2O.$$
$$\text{Aurin.} \qquad\qquad \text{Red corallin.}$$

Red corallin comes into commerce as peonin. "Spirit-soluble" if in the free state, and "water-soluble" as the ammonium salt.

In the first case it forms lumps with a metallic lustre, in the latter a brown-red porous mass. It dissolves in concentrated sulphuric acid with a yellow colour.

The red solution of the ammonium salt is unaltered by alkalies, and is precipitated yellow by acids. Metallic salts, as basic acetate of lead, acetate of alumina, tin chloride, etc., produce orange-red or yellow precipitates.

Application.—Neither the red nor the yellow corallin is used in dyeing, as the yellow or red orange shades produced are very unstable. It can be washed with water, but will not stand the action of either acids, alkalies, or light.

Somewhat better results are obtained in printing.

For printing on wool, the colour is thickened with gum water or glycerin, and a little magnesia or zinc oxide added, to overcome the action of acids. The colours can be steamed.

For printing on silk, the colour is prepared by dissolving the tin lake in oxalic acid.

For calico-printing, the same process can be used as for wool, only albumen is substituted for the gum-water.

Detection on the fibre.—The reaction with dilute acids (yellow) is characteristic. Alkalies and ammonia produce red solutions, which do not fluoresce. Chloride of lime decolourises immediately.

Rosophenolin is obtained by the action of alcoholic ammonia or aurin in presence of phenol or benzoic acid.

(d) PHTHALEÏNS.

Crude materials.—The crude materials for the preparation of the phthaleïns are the phenols and phthalic acid or its anhydride.

Reactions of phthalic acid and phenols.—If a phenol is

heated with phthalic anhydride, a combination takes place, with elimination of water. At a moderate temperature, two molecules of the phenol unite with one molecule of phthalic anhydride, forming a "phthaleïn." At higher temperatures with one molecule of each, a derivative of anthracene is formed. From phenol, C_6H_5OH, phthalic anhydride and sulphuric acid are produced, according to the temperature.

$$(1)\ C_6H_4\!\!\begin{array}{c}\diagup C|O|\diagdown\\ \diagdown CO \diagup\end{array}\!\!O + 2C_6H_4\cdot|H|\cdot OH =$$

Phthalic anhydride.   Phenol.

$$C_6H_4-C\begin{array}{c}\diagup C_6H_4\cdot OH\\ \diagdown C_6H_4\cdot OH\\ \diagdown CO \diagdown O\end{array} + H_2O.$$

Phenol phthaleïn.

$$(2)\,C_6H_4\!\!\begin{array}{c}\diagup CO\diagdown\\ \diagdown CO \diagup\end{array}\!\!O + C_6H_5OH = C_6H_4\!\!\begin{array}{c}\diagup CO\diagdown\\ \diagdown CO \diagup\end{array}\!\!C_6H_3OH + H_2O.$$

Phthalic anhydride. Phenol. Oxyanthraquinon.

The formula of phenol phthaleïn is better expressed as—

$$C\left\{\begin{array}{l}C_6H_4\cdot OH\\ C_6H_4\cdot OH\\ C_6H_4\cdot CO\\ \qquad\ \ |\\ \underline{\qquad\qquad}O\end{array}\right.$$

The phthaleïns are, therefore, along with rosaniline and rosolic acids, derivatives of triphenylmethane, $C(C_6H_5)_3H$.

The relation between these groups of colouring matters is shown by the constitutional formulæ:—

$$C\left\{\begin{array}{l}C_6H_4\cdot NH_2\\ C_6H_4\cdot NH_2\\ C_6H_4\cdot NH.\\ \underline{\qquad\quad\ |}\end{array}\right.\qquad C\left\{\begin{array}{l}C_6H_4\cdot OH\\ C_6H_4\cdot CH\\ C_6H_4\cdot O.\\ \underline{\qquad\quad\ |}\end{array}\right.\qquad C\left\{\begin{array}{l}C_6H_4\cdot OH\\ C_6H_4\cdot OH\\ C_6H_4\cdot CO.\\ \qquad\quad\ |\\ \underline{\qquad\qquad}O.\end{array}\right.$$

Pararosaniline.   Aurin.   Phenol phthaleïn.

A confirmation of this formula is obtained as follows :—
Diphenyl phthalid—

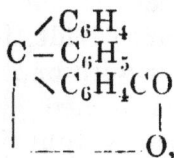

$$C - \begin{cases} C_6H_4 \\ C_6H_5 \\ C_6H_4CO \end{cases}$$
$$\underline{\qquad\qquad} O,$$

gives successively—

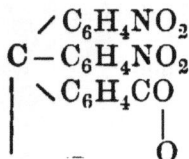

$$C - \begin{cases} C_6H_4NO_2 \\ C_6H_4NO_2 \\ C_6H_4CO \end{cases}$$
$$\underline{\qquad\qquad} O$$

Dinitrodiphenyl
phthalid.

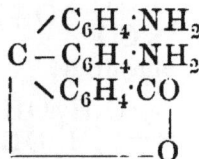

$$C - \begin{cases} C_6H_4 \cdot NH_2 \\ C_6H_4 \cdot NH_2 \\ C_6H_4 \cdot CO \end{cases}$$
$$\underline{\qquad\qquad} O$$

Diamidodiphenyl
phthalid.

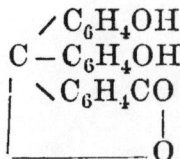

$$C - \begin{cases} C_6H_4OH \\ C_6H_4OH \\ C_6H_4CO \end{cases}$$
$$\underline{\qquad\qquad} O$$

Phenol phthaleïn.

if it is warmed with nitric acid, the product reduced with zinc and hydrochloric acid, and the dilute aqueous solution heated with nitrite of soda and sulphuric acid.

The phthaleïns, like the rosanilines and rosolic acids, give leuco compounds on reduction, which are known as phthalins.

Phenol phthalin has the formula—

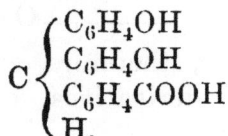

$$C \begin{cases} C_6H_4OH \\ C_6H_4OH \\ C_6H_4COOH \\ H. \end{cases}$$

The phthaleïns of resorcin and pyrogallol are the only ones of commercial importance.

Fluoresceïn.

Fluoresceïn, $C_{20}H_{12}O_5$.—All the colouring matters known as eosins are derived from fluoresceïn, an anhydride of resorcin phthaleïn. To prepare it, the molecular proportions of phthalic anhydride and resorcin are heated to 195°–200° till no more steam is evolved, and the mass has become solid. The melt is cooled, and pulverised, and serves as crude fluoresceïn for the preparation of the eosins. Fluoresceïn is formed according to the equation :—

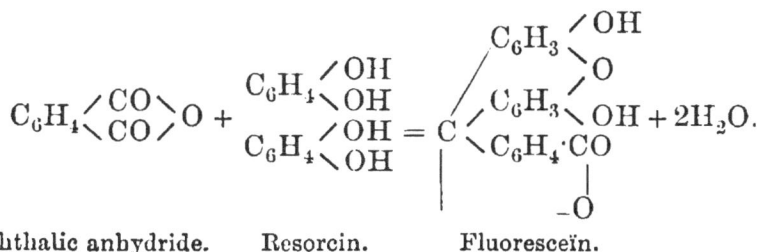

$$C_6H_4\!\!<\!\!{CO\atop CO}\!\!>\!\!O + {C_6H_4\!\!<\!\!{OH\atop OH}\atop C_6H_4\!\!<\!\!{OH\atop OH}} = C\!\!<\!\!{{C_6H_3\!\!<\!\!{OH\atop O}\atop C_6H_3\!\!<\!\!{\atop OH}}\atop C_6H_4\!\cdot\! CO}\Big|_{-O} + 2H_2O.$$

Phthalic anhydride.　　Resorcin.　　Fluoresceïn.

Pure fluoresceïn forms a yellowish-red crystalline powder, which is sparingly soluble in cold water, with a yellow colour. It is a weak acid, being displaced in its salts by acetic acid.

Its alkaline solution possesses a bright-green fluorescence, which is so intense that 1 part dissolved in alkali, and diluted with 2,000,000 parts of water, still shows a fluorescence. Whole rivers may be coloured for a time with a single kilogram. This property has been made use of for determining the course of rivers which sink for a time into the ground. Thus it has been proved by this means that there is an underground connection between the Danube and the Ach, a small river which flows into the Lake of Constance.

By warming with soda and zinc powder, it is reduced to fluorescin—

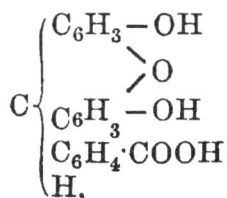

$$C\begin{cases}C_6H_3-OH\\ \quad\searrow O\\ \quad\nearrow\\ C_6H_3-OH\\ C_6H_4\!\cdot\!COOH\\ H,\end{cases}$$

the alkaline solution of which is gradually oxidised in the air to fluorescein.

Fluorescein dyes silk and wool yellow, but is seldom used in dyeing, as the colours are not fast. Its sodium salt comes into commerce under the name of *Uranin*, and finds a limited application in woollen printing. It forms insoluble lakes with lead, silver, etc., which being non-poisonous, may be used for colouring toys, india-rubber goods, etc.

From fluorescein, two series of commercial dye-stuffs are derived. The first comprises the unsubstituted ethers of fluorescein, and the other the nitro and halogen derivatives and their ethers, which are generally called eosins.

Chrysolin.

Benzyl fluorescein, $C_{20}H_{10}O_3(OC_7H_7)OH$, is produced by heating phthalic acid, benzyl resorcin ($C_6H_4 \cdot OC_7H_7 \cdot OH$) and sulphuric acid. Its sodium salt is the chrysolin of commerce.

Chrysolin is a reddish-brown powder; the larger pieces possess a greenish reflex. It dissolves in water and alcohol with a green fluorescence. From the solutions, acids precipitate free benzyl fluorescein.

With stannous chloride and with lead-salts, it gives beautifully coloured lakes.

On silk and wool it yields a fast yellow, very similar to that produced with turmeric. It is used in cotton-dyeing for topping quercitron-yellow, the quercitron itself acting as a mordant for benzyl fluorescein.

Eosins.

Tetrabromfluorescein, Eosin yellowish, Eosin A, Eosin GGf, Soluble Eosin, Eosin B, Eosin A extra, $C_{20}H_8Br_4O_5$.—If a dilute solution of fluorescein in soda is mixed with the calculated quantity of bromine dissolved in soda, and dilute hydrochloric acid is then added, free fluorescein is first precipitated, but this unites immediately with the liberated bromine. The precipitate is filtered, pressed, neutralised with soda or potash, and the solution evaporated to dryness.

The product comes into commerce as eosin yellow shade, soluble eosin, or eosin J. The pure re-crystallised potash salt has the composition $C_{20}H_6Br_4O_5K_2 + 6H_2O$. It forms a red powder, or red crystals with yellowish-green reflex. One part dissolves completely in 2 to 3 parts of water; it is not easily soluble in absolute alcohol.

Its dilute aqueous solutions are rose-coloured, with an intense green fluorescence; alcohol renders the fluorescence still more intense. Free tetrabromfluorescein is nearly insoluble in water, while its reddish-yellow alcoholic solution shows no fluorescence. It is a pretty strong acid, its salts not being decomposed by acetic acid.

It combines with metallic oxides, producing sparingly soluble or insoluble lakes. They are prepared by mixing aqueous solutions of metallic salts and eosins, when they separate as amorphous, bright-coloured precipitates. Silver and lead salts give red, aluminium, zinc, tin, cobalt, iron, manganese and bismuth salts give reddish-yellow lakes.

Eosin orange and *Eosin 5G* are mixtures of dibromfluorescein and tetrabromfluorescein.

Tetraiodfluorescein, $C_{20}H_8I_4O_5$, is prepared in a similar manner to tetrabromfluorescein. Its alkaline salts are sold as *eosin blue shade* (soluble in water), *erythrosin, pyrosin R, soluble primrose, Dianthin B, Rose B soluble in water, Eosin J*. Its aqueous solutions are not fluorescent, otherwise it resembles eosin J.

Aureosin J is a chlorinated fluorescein.

Bromnitrofluorescein are produced by the action of dilute nitric acid on tetrabromfluorescein. Their salts are known in commerce as *eosin BN* or *safrosin*.

A mixture of bromnitrofluorescein with di- and tetra-nitrofluorescein is known as *lutécienne*. *Rubeosin* is a nitrochlorofluorescein obtained by the action of nitric acid on aureosin.

The alcohol-soluble eosins are substitution products of the methyl and ethyl ethers of fluorescein.

The methyl and ethyl ethers of tetrabromfluorescein are produced by heating tetrabromfluorescein with wood spirit, or ethyl alcohol, and sulphuric acid. They may also be produced by heating eosin with methyl or ethyl bromide. Their potassium salts—

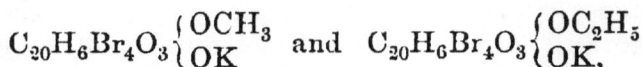

$$C_{20}H_6Br_4O_3 \begin{cases} OCH_3 \\ OK \end{cases} \text{ and } C_{20}H_6Br_4O_3 \begin{cases} OC_2H_5 \\ OK, \end{cases}$$

are commercial products under the names, *eosin* soluble in spirit, *ethyl eosin, methyl eosin, spirit primrose, rose JB soluble in spirit, primrose, eosin BB.*

They are sparingly soluble in water, but dissolve easily in alcohol of 50 per cent., while they are insoluble in absolute alcohol. The dilute solutions possess a beautiful fluorescence.

These eosins contain the substituted halogen atoms in the resorcin rest. Eosins may also be prepared containing the halogen atoms in the phthalic-acid rest, thus dichlorphthalic acid acts on resorcin, producing dichlor-fluorescein :—

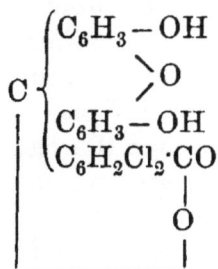

$$C \begin{cases} C_6H_3-OH \\ \quad \diagdown O \\ \quad \diagup \\ C_6H_3-OH \\ C_6H_2Cl_2 \cdot CO \\ \qquad \qquad | \\ \qquad \qquad O \\ \qquad \qquad | \end{cases}$$

Rose Bengale is the sodium salt of tretraioddichlor-fluorescein :—

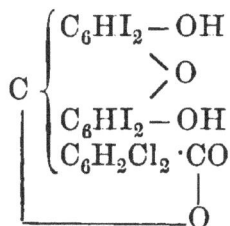

$$C \begin{cases} C_6HI_2-OH \\ \quad\quad > O \\ C_6HI_2-OH \\ C_6H_2Cl_2 \cdot CO \\ \quad\quad\quad | \\ \overline{\quad\quad\quad\quad O} \end{cases}$$

Phloxiu TA, *Phloxin, Eryturosin* B, is the potassium salt of tetrabromtetrachlorfluorescein, $C_{20}H_2Cl_4Br_4O_5Na_2$, and *cyanosine* B, is the sodium salt of the ethyl ether of phloxin, $C_{20}H_2Cl_4Br_4O_5C_2H_5Na$.

Phloxin P, Phloxin is the potassium salt of tetrabrom-dichlorfluorescein, $C_{20}H_4Cl_2Br_4O_5K_2$. *Cyanosin* is the potassium salt of the methylether of tetrabromdichlor-fluorescein.

Reactions of the eosins.—The fluorescence of the eosins is most intense in alcoholic solutions. It is the brightest in the "alcohol-soluble eosins," then comes eosin G. Safrosine fluoresces very little, somewhat more in presence of ammonia, while eosin B does not fluoresce at all in aqueous solution and very little in alcoholic solution.

The eosins dissolve in cold concentrated sulphuric acid with a yellow colour, which generally becomes darker on heating. Eosin G and BN are turned dark-red, and, on adding water, a new colouring matter separates in dark flakes. On heating eosin B with sulphuric acid, iodine separates out.

Hydrochloric acid separates from eosin solutions the free fluoresceins. Eosin G gives a yellow, eosin B an orange-red, safrosin a yellow-brown, and spirit-soluble eosin a beautiful red, precipitate.

Metallic salts give coloured precipitates.

Chloride of lime decolourises the solutions on heating.

With the exception of eosin B, all eosins may be recognised by their fluorescence, especially in presence of ammonia. The reactions with sulphuric and hydrochloric acids are also characteristic.

The recognition of eosin B, and the separation of the colouring matters, is best carried out by the reactions with zinc powder and ammonia. The iodine is thus removed from eosin B, and the fluorescein produced is reduced to colourless fluorescin, which on oxidation in the air yields fluorescein.

Eosin G does not lose any bromine, but is reduced to colourless tetrabromfluorescin, which yields the original tetrabromfluorescein.

Eosin BN is reduced to colourless fluorescin, and the nitro groups are simultaneously reduced; on oxidation a non-fluorescent cherry-red solution is produced.

The "spirit-soluble" methyl and ethyl eosins react in the same way as eosin G.

The oxidation of the reduced solutions in the air takes place immediately only with safrosin; the other colouring matters require to stand a longer time. Characteristic colourations are obtained by proceeding as follows :—

A small quantity of the very dilute aqueous solution of the colouring matter is mixed with a few drops of ammonia and zinc powder till completely decolourised; it is then filtered, and the filtrate boiled till the excess of ammonia is removed, and a precipitate of zinc hydrate is produced. Then hydrochloric acid is added till the precipitate re-dissolves, when the liquid is treated with excess of ammonia.

The following table of reactions will be understood without further explanation :—

	Original dilute solution.	Filtrate immediately after boiling with Zn. and NH_3.	After oxidation by boiling, etc.
Eosin G.	Cherry-red, yellow-green fluorescence.	Nearly colourless.	Cherry-red yellowish-green fluorescence.
Eosin B.	Cherry-red, without fluorescence.	Nearly colourless.	Yellow, with strong green fluorescence.
Eosin BN.	Cherry-red, yellowish-green fluorescence.	Cherry-red, non-fluorescent.	Cherry-red, non-fluorescent.
Spirit-soluble Eosin.	Cherry-red, strong green fluorescence.	Nearly colourless.	Cherry-red, yellow-green fluorescence.

The eosins also give characteristic colour reactions, when a small quantity is boiled with strong potash lye (containing 20 to 40 per cent. of solid KOH).

Eosin G gives in the cold an orange-red solution, which on boiling becomes purple-red, violet, and pure blue, with a *strong* green fluorescence. The colour and fluorescence are unaltered on dilution, if the boiling has been continued long enough.

Eosin B, on boiling with potash, first turns purple-red and then blue-violet, with a weak, green fluorescence. On dilution the solution becomes purple-red.

Eosin BN becomes lighter and yellower with potash, and on boiling turns olive-green, without fluorescence.

Spirit-soluble eosins are insoluble in the concentrated lye; on boiling the same reactions as with eosin G gradually take place.

Application.—Eosins yield bright shades on silk, wool, and cotton, comprising all the shades from a reddish-

orange to a cherry-red and purple. The bluest shades are
produced by rose Bengale; then come safrosin, phloxin,
and eosin B, and the methyl and ethyl eosins. The
yellowest is eosin G.

Safrosin is very bright on wool, but is surpassed on
silk by the other bluish eosins. The spirit-soluble eosins
are brighter than those soluble in water, but they are not
often used, as they are more expensive. It is necessary
to dissolve them for use in methylated spirits, and the
colours produced have a strong fluorescence, which is
seldom desirable.

Silk is dyed in boiled-off liquor, and brightened with
acetic or tartaric acid.

Wool is dyed with alum or acetic acid. A good result
is obtained by the sulphur mordant, as applied in dyeing
methyl-green.

Cotton is mordanted with alumina or stannate of soda
for yellow shades, with lead-salts for blue shades. In
mordanting with alumina, the cotton is first passed
through a hot, strong solution of soap, and, after squeezing
through a solution of aluminium acetate at about 12° Tw.
Oleate, palmitate or stearate of alumina is thus precipitated on the fibre, which now takes up the dye very readily.

Eosins are often used mixed with other colouring
matters, such as red, yellow, and violet, and mixtures of
this kind are sometimes sold as commercial products.
Nopaline, or Imperial-red, contains Martius-yellow, while
coccine contains aurantia.

Safrosin, when mixed with yellow colouring matters,
is sold as cochineal substitute, and yields a scarlet very
similar in appearance to cochineal-scarlet.

Detection on the fibre.—Warm water removes a trace of the
colouring matter, especially if a little ammonia is added,
from goods dyed with eosins soluble in water. Spirit-
soluble eosins are not affected by water, but alcohol removes
them, and leaves the eosins soluble in water on the fibre.

Concentrated potash solution gives the reactions described above. Hydrochloric acid or acidulated stannous chloride decolourise or turn the fabric yellow. Sulphuric acid turns them yellow, chloride of lime decolourises on warming.

The colours produced on cotton are not so fast as on wool or silk. They are not fast to light.

Rhodamine.

By heating phthalic anhydride with diethylamido-phenol, $C_6H_4 \begin{cases} N (C_2H_5)_2 \\ OH \end{cases}$ a phthaleïn is formed which possesses simultaneously basic and acid properties and the hydrochloride of which forms the commercial rhodamine. Its constitution is represented by the formula—

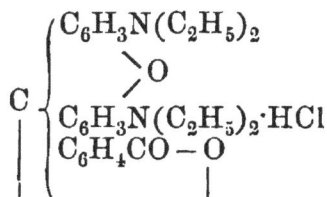

$$C \begin{cases} C_6H_3N(C_2H_5)_2 \\ \quad \diagdown O \\ \quad \diagup \\ C_6H_3N(C_2H_5)_2 \cdot HCl \\ C_6H_4CO - O \end{cases}$$

Commercial rhodamine forms a reddish-coloured powder easily soluble in water with a magenta colour. The dilute aqueous solution shows a strong fluorescence. The free base is precipitated from the aqueous solution by caustic soda, in red flakes, soluble in ether. With stannous chloride rhodamine gives a brilliant scarlet precipitate, which in a fine state of division shows a remarkable fluorescence; the reflected light is scarlet, while the transmitted light is of a bright-blue colour.

Application.—Rhodamine produces on wool and silk an exceedingly bright pink with yellow fluorescence, which is especially valuable on account of its comparative fastness to light. Cotton is previously mordanted with tannin and tartar emetic.

R

Galleïn.

Galleïn, $C_{20}H_{10}O_7$, is obtained by heating phthalic anhydride and pyrogallic acid to 190°–200° till a solid mass is formed. The melt is boiled with water, dissolved in sodium carbonate, filtered, and the colouring matter precipitated with an acid.

A combination of phthalic anhydride and pyrogallic acid takes place with elimination of water, but at the same time an oxidation takes place :—

$$C_6H_4 {<}{CO \atop CO}{>}O + 2C_6H_3 {\begin{cases} OH \\ OH \\ OH \end{cases}} + O = C {\begin{cases} C_6H_2OH{\cdot}O \\ \qquad {>}O \\ C_6H_2OH{\cdot}O \\ C_6H_4{\cdot}CO{\cdot}O \end{cases}}$$

Phthalic anhydride. Pyrogallic acid.

Galleïn.

Galleïn forms either a brown-red powder or green metallic-looking crystals, which are almost insoluble in water, but soluble in alcohol. It is a weak acid, forming a blue solution with alkalies. Its alkaline solution assumes a dirty colour on standing. It dissolves in ammonia with a violet colour, which solution gives violet precipitates with metallic salts.

Galleïn yields on wool, mordanted with bichromate and tartar, dark shades of violet which are characterised by their beautiful "bloom." On mordanted cotton, it produces fast violet shades. Lead mordants give a fine grey-violet.

In calico-printing, it has been superseded by the fast alizarin-violet. It may be fixed, like coeruleïn, with chromium acetate.

Galleïn is most important as a raw material for the preparation of coeruleïn.

Coeruleïn.

Galleïn is heated to 200° with twenty times its weight of strong sulphuric acid, when the red colour gradually changes to a brownish-green. The mixture is cooled and poured into water, and the black precipitate washed and dried.

The formation of coeruleïn is expressed by the equation—

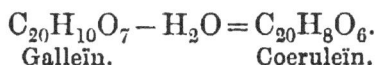

$$C_{20}H_{10}O_7 - H_2O = C_{20}H_8O_6.$$
Galleïn.   Coeruleïn.

Its constitution has not been determined with certainty, but, according to Buchka, it is a derivative of phenyl anthracene—

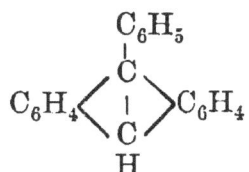

$$C_6H_4 \left\langle \begin{array}{c} C_6H_5 \\ | \\ C \\ | \\ C \\ | \\ H \end{array} \right\rangle C_6H_4$$

and thus belongs to the anthracene colours.

Properties.—Dry coeruleïn forms an amorphous, almost black mass. It is nearly insoluble in water, alcohol and ether. The best solvent is acetic acid. Commercial coeruleïn forms a thick, dark-coloured paste. It dissolves in alkalies with a beautiful green colour. This solution gives with mordants very stable lakes. On warming with ammonia and zinc powder, a brown-red solution is formed, containing coerulin, the phthalin of coeruleïn. The solution is oxidised in the air, the original green colour being restored. It may be used for dyeing as a coeruleïn vat, and acts in the same way as the indigo vat.

Bisulphite compound. — On stirring the commercial coeruleïn paste with a concentrated solution of sodium bisulphite, $NaHSO_3$, a colourless compound of coeruleïn and bisulphite is formed, which can be removed from the unaltered coeruleïn by extracting with cold alcohol. The best proportions are one molecule of coeruleïn to two molecules of bisulphite.

Sulphurous acid or ammonium bisulphite acts in a similar manner.

The coeruleïn sulphite does not contain a reduced colouring matter, but is a double compound, analogous to those formed with aldehydes and ketones. It is colourless, and soluble in water and alcohol. On boiling, the greenish-yellow solution turns green, and becomes alkaline, sulphurous acid being ' evolved. Acids and alkalies decompose coeruleïn sulphite in the cold, with evolution of sulphurous acid.

Coeruleïn sulphite comes into commerce as coeruleïn S. It is easier to dye with than the paste. It forms a black powder, which dissolves in water with a dull-green colour, the solution giving all the reactions of the solid compound. The pure compound may be obtained by crystallising coeruleïn from alcohol.

Application.—Coeruleïn is used in large quantities for dyeing or printing cotton fabrics. With chromium mordants it gives a very fast, but dull olive-green, which, however, in point of fastness equals the alizarin colours. Other green dyes, such as chrome-green (Guignet's green) and aniline-green, are brighter, but not nearly so fast.

For printing with coeruleïn S, the latter is mixed with chromium acetate, and thickened, and fixed by steaming. If insoluble coeruleïn is used, it is previously rendered soluble by the addition of sodium bisulphite.

In wool-dyeing, coeruleïn also gives excellent results. It may be used for this purpose either in the form of the bisulphite compound or as the paste. The most effective mordant for wool is bichromate of potash (3%), and tartar ($2\frac{1}{2}$%). The shades obtained resemble woaded greens. Coeruleïn can also be used along with many of the natural colouring matters and with the alizarins, by which means a large variety of fast shades can be produced.

Detection on the fibre.—The colour is not removed by

boiling soap or caustic alkalies. Concentrated hydrochloric acid darkens the colour. The most characteristic
reaction is the action of a warm, acid solution of stannous
chloride, which turns the fibre brown-red, coerulin being
formed. On washing with water, or, better still, with
very dilute chloride of lime solution, the original colour
is restored.

Quinoline-yellow.

Quinoline-yellow is the sodium salt of quinophthalondi-
sulphonic acid.

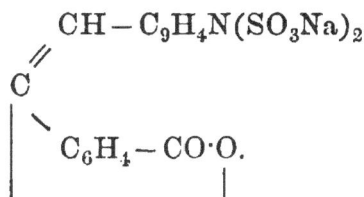

$$CH - C_9H_4N(SO_3Na)_2$$
$$C$$
$$C_6H_4 - CO \cdot O.$$

It is obtained by treating quinaldine $C_{10}H_9N$ with
phthalic anhydride in presence of zinc chloride, and sulphonating the product obtained.

It forms a bright-yellow powder, which dissolves easily
in water with an intense yellow colour, which is unaltered by dilute sulphuric acid, but is turned somewhat
darker by ammonia.

Application. — Quinoline-yellow is not applicable to
cotton. It is dyed on wool and silk in an acid-bath
(H_2SO_4), and yields very pure shades of yellow, which
somewhat resemble those obtained with picric acid. It
stands light fairly well, only becoming a trifle lighter after
a month's exposure, in which it differs materially from
picric acid, which becomes orange.

Quinoline-blue, $C_{29}H_{35}N_2I$ is obtained by treating equal
molecules of quinoline and lepidine with amyl iodide and
subjecting the product obtained to the action of caustic
alkalis.

It forms green crystals, which dissolve in boiling water with a violet-blue colour, and is used for sensitising photographic plates.

Quinoline-red, $C_{26}H_{19}N_2Cl$ is formed by the action of benzotrichloride on a mixture of quinaldine and isoquinoline in presence of zinc chloride. It forms dark brown-red needles with a bronze reflex, which dissolve in hot water with a red colour. It is used like quinoline-blue for sensitising photographic plates.

III. THE AZO DYES.

THE azo dyes form a well-defined and well-studied group of the coal-tar colours. Their chemical constitution has been clearly proved in every case as far as possible in accordance with our present knowledge of chemistry. All the azo dyes may be prepared according to one general method, viz., by acting on diazo compounds with phenoles or amines of the aromatic series.

Diazo compounds are formed (as stated on several previous occasions) when primary amines of the aromatic series are treated with nitrous acid. In order to be able to understand the manufacture of the azo dyes properly, it will be necessary to know something of the properties of these compounds.

If nitrous fumes, obtained by heating starch with nitric acid possessing a specific gravity of 1·30 to 1·35, are passed into an aqueous solution of nitrate of aniline, the following equation is fulfilled :—

$$C_6H_5NH_2 \cdot HNO_3 + HNO_2 = C_6H_5 - N:N \cdot NO_3 + 2H_2O.$$

Nitrate of aniline.   Nitrate of diazobenzene.

A similar reaction takes place if a solution of aniline hydrochloride is treated first with hydrochloric acid and then with sodium nitrite, in the proportions indicated in the equation :—

$$C_6H_5NH_2 \cdot HCl + HCl + NaNO_2 =$$

Aniline hydrochloride.

$$C_6H_5NH:N \cdot Cl + NaCl + 2H_2O.$$

Diazobenzene chloride.

If nitrous fumes are passed into an alcoholic solution of aniline, a diazo compound is likewise formed, which is known as diazoamidobenzene, and has the formula $C_6H_5N:N\cdot NH\cdot C_6H_5$. The formation of this compound is easily explained by supposing the equation to be fulfilled in two phases :—

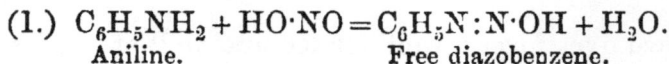

(1.)　$C_6H_5NH_2 + HO\cdot NO = C_6H_5N:N\cdot OH + H_2O$.
　　　　Aniline.　　　　　　　　　　Free diazobenzene.

The diazobenzene acts in the nascent state on a second molecule of aniline, thus :—

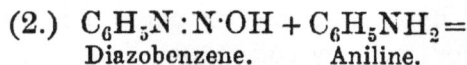

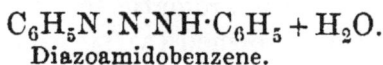

(2.)　$C_6H_5N:N\cdot OH + C_6H_5NH_2 =$
　　　Diazobenzene.　　　　Aniline.

$C_6H_5N:N\cdot NH\cdot C_6H_5 + H_2O$.
Diazoamidobenzene.

The formulæ given in these equations for—

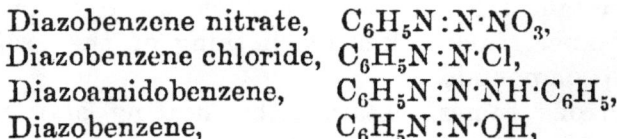

Diazobenzene nitrate,	$C_6H_5N:N\cdot NO_3$,
Diazobenzene chloride,	$C_6H_5N:N\cdot Cl$,
Diazoamidobenzene,	$C_6H_5N:N\cdot NH\cdot C_6H_5$,
Diazobenzene,	$C_6H_5N:N\cdot OH$,

are typical as representatives of this group, which latter can therefore be characterised in the following manner :—

" Diazo compounds are bodies which contain the diatomic group $-N:N-$, which is bound on one side to a carbon atom, the other bond being saturated with oxygen, nitrogen, chlorine, bromine, etc."

Most of the diazo compounds crystallize well and are colourless, with the exception of the diazoamido compounds, which are either yellow or red. When heated, they detonate with extreme violence. The diazo compounds decompose very easily. Thus free diazobenzene decomposes immediately when separated from its salts. The stability of the compounds is, however, increased by the introduction of electro-negative groups into the benzene ring. Thus diazonitrophenol can be kept in the free state without undergoing any change.

If aqueous solutions of the salts of diazo compounds are heated, decomposition ensues along with the evolution of nitrogen gas and formation of resinous and other products. If, however, the solution is acidulated before boiling, a phenol is the chief product of the reaction ; e,g. :—

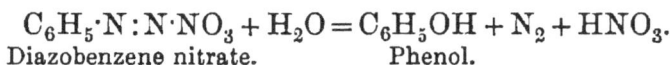

$$C_6H_5 \cdot N : N \cdot NO_3 + H_2O = C_6H_5OH + N_2 + HNO_3.$$

Diazobenzene nitrate.   Phenol.

The fact that the diazo compounds are so easily decomposed, along with the great danger of keeping large quantities of such explosive substances, would render their application in the manufacture of coal-tar colours prohibitive, should it be necessary to prepare them in the dry state. This can, however, be avoided by preparing them in very dilute solutions, from which the azo bodies can be separated directly without isolating the diazo compounds.

The *azo compounds* also contain the diatomic group $-N:N-$, which is, however, bound on either side to a carbon atom. The simplest representative of this class of bodies is known as azobenzene, and has the formula—

$$C_6H_5N : NC_6H_5.$$

Azobenzene and its homologues can be obtained by reduction of the nitro derivatives of the corresponding hydrocarbons. The most suitable reducing agents are sodium amalgam, zinc and alcoholic potash, or an alkaline solution of stannous hydrate. In the case of nitrobenzene, the reaction takes place according to the equation—

$$2C_6H_6NO_2 + 4H_2 = C_6H_5 \cdot N : N \cdot C_6H_5 + 4H_2O.$$

Nitrobenzene.   Azobenzene.

The simple azo compounds are for the most part brightly-coloured bodies; but they are no colouring matters, since they do not possess the property of combining with either acids or bases.

The azo dyes are amido- or hydroxyl-derivatives of the simple azo compounds, and are distinguished as amidoazo and oxyazo dyes. From azobenzene, for instance, the following representatives of these two groups may be derived :—

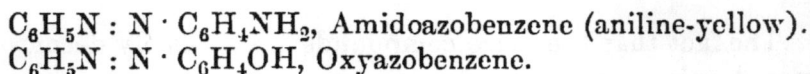

$C_6H_5N : N \cdot C_6H_4NH_2$, Amidoazobenzene (aniline-yellow).
$C_6H_5N : N \cdot C_6H_4OH$, Oxyazobenzene.

For dyeing, the amidoazo dyes can either be used as such, or in the form of their sulphonic acids, while the oxyazo dyes nearly always contain sulpho groups. It is for this reason that the azo dyes have been divided below into *three* groups.

The azo dyes are never prepared in practice from the simple azo compounds, although this would be possible in one or two cases. Thus, amidoazobenzene can be obtained by nitrating azobenzene, which yields nitroazobenzene, $C_6H_5N:NC_6H_4 \cdot NO_2$, and this, when reduced, yields amidoazobenzene. The general method mentioned above, *i.e.*, allowing diazo compounds to act on phenols or amines, has not only the advantage of being easily carried out, but also allows a larger number of combinations to be made.

The following example, the preparation of dimethyl-amidoazobenzene, may serve to illustrate the reaction :—

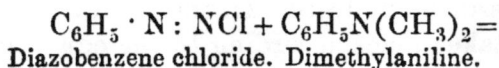

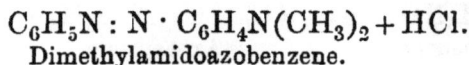

$$C_6H_5 \cdot N : NCl + C_6H_5N(CH_3)_2 =$$
Diazobenzene chloride. Dimethylaniline.

$$C_6H_5N : N \cdot C_6H_4N(CH_3)_2 + HCl.$$
Dimethylamidoazobenzene.

If aniline and diazobenzene hydrochloride are allowed to act on each other in a similar manner, the reaction is somewhat different, diazoamidobenzene being first formed :—

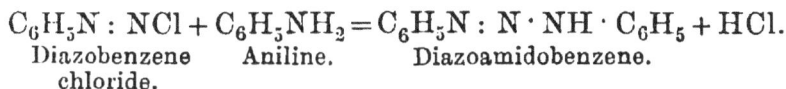

$$C_6H_5N : NCl + C_6H_5NH_2 = C_6H_5N : N \cdot NH \cdot C_6H_5 + HCl.$$

Diazobenzene   Aniline.   Diazoamidobenzene.
chloride.

This latter product soon passes over, however, into the isomeric amidoazobenzene, especially in the presence of a small quantity of aniline hydrochloride :—

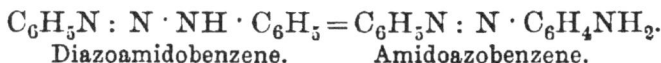

$$C_6H_5N : N \cdot NH \cdot C_6H_5 = C_6H_5N : N \cdot C_6H_4NH_2.$$

Diazoamidobenzene.   Amidoazobenzene.

In all cases hitherto noticed of amines or phenoles combining with diazo compounds, the amido or hydroxyl group invariably assumes the para position with respect to the –N:N– group.

Of the general reactions of the azo compounds, we will only mention here their behaviour towards energetic reducing agents. The latter split up the azo compound into two halves, and at the same time transform the –N : N– group into two amido groups; e.g. :—

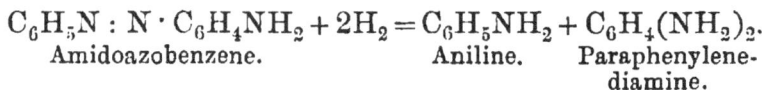

$$C_6H_5N : N \cdot C_6H_4NH_2 + 2H_2 = C_6H_5NH_2 + C_6H_4(NH_2)_2.$$

Amidoazobenzene.   Aniline.   Paraphenylenediamine.

√(a) AMIDOAZO DYES.

Only three technical products belong to this group—viz., aniline-yellow, chrysoidine and phenylene-brown. The colour-bases contained in them are all derived from azo-benzene, $C_6H_5 -N: N- C_6H_5$, and they form the following series :—

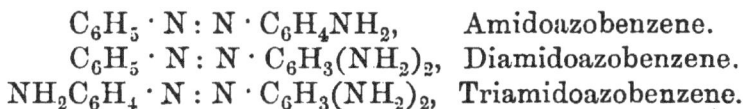

$C_6H_5 \cdot N : N \cdot C_6H_4NH_2$,   Amidoazobenzene.
$C_6H_5 \cdot N : N \cdot C_6H_3(NH_2)_2$,   Diamidoazobenzene.
$NH_2C_6H_4 \cdot N : N \cdot C_6H_3(NH_2)_2$,   Triamidoazobenzene.

The affinity of these bases to acids, and at the same time their fastness on the fibre, is proportionate to the number of amido groups they contain; thus, phenylene-brown is faster than chrysoidine, and this again is faster than aniline-yellow.

It is a characteristic of these dyes that the aqueous solutions of their hydrochlorides are precipitated by alkalies and ammonia, and reddened by strong acids.

ᴸ Aniline-yellow.

Aniline-yellow, or amidoazobenzene hydrochloride, $C_6H_5N : N \cdot C_6H_4NH_2HCl$, can be obtained by mixing dilute aqueous solutions of diazobenzene chloride and aniline.

In the pure state it forms bluish-violet, lustrous needles, which dissolve in acidulated water with a fine red colour. If the solution is heated, the salt is decomposed and the free base is precipitated. Free amidoazobenzene can be completely precipitated from solutions of its salts by ammonia. In the pure state it forms yellow crystals, which melt at 127·5°, and can be volatilised without decomposition. It is insoluble in water, but dissolves in alcohol.

In acid solutions of aniline-yellow, silk is dyed red, the salt itself being assimilated by the fibre. But on washing with water, the salt is decomposed, and only the free base remains on the fibre, to which latter it imparts a yellow colour.

Aniline-yellow is at present not used on the large scale, since it is not at all fast and easily volatilises when steamed.

✓ Chrysoidine.

Chrysoidine, or diamidoazobenzene hydrochloride, $C_6H_5N : N \cdot C_6H_3(NH_2)_2 \cdot HCl$, is formed when an aqueous solution of metaphenylenediamine is poured into a very dilute solution of diazobenzene chloride:—

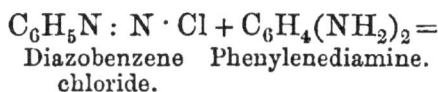

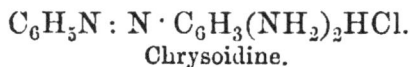

$$C_6H_5N : N \cdot Cl + C_6H_4(NH_2)_2 =$$

Diazobenzene   Phenylenediamine.
chloride.

$$C_6H_5N : N \cdot C_6H_3(NH_2)_2HCl.$$

Chrysoidine.

Chrysoidine separates out as a sparingly soluble precipitate.

Metaphenylenediamine may be obtained in a comparatively pure state by first converting benzene by means of concentrated nitric and sulphuric acids into dinitrobenzene, $C_6H_4(NO_2)_2$, and reducing this compound with iron and hydrochloric acid.

The chrysoidine of commerce usually consists of dark violet crystals, having a metallic reflex, which dissolve easily and without decomposition in boiling water and in absolute alcohol. From the orange-coloured solution ammonia or caustic soda throw down a bright-yellow precipitate, which consists of the free colour-base. Chrysoidine contains two amido groups, and is therefore also capable of combining with two equivalents of an acid. The salts formed are, however, not at all stable. The fact that solutions of chrysoidine are turned yellow by excess of hydrochloric acid is due to the formation of these abnormal salts. In very thin layers, these solutions appear crimson. Concentrated sulphuric acid dissolves chrysoidine with a yellowish-brown colour.

Stannous chloride decolourises in the cold. Basic lead acetate produces an orange precipitate.

Cardinal is a mixture of magenta and chrysoidine. A mixture of safranine and chrysoidine is sometimes sold as *cotton scarlet* (Schulz & Julius).

Application.—Silk is dyed in a neutral soap-bath, wool

in pure water. Cotton is mordanted with tannin and tartar emetic. The solution of the dye should be filtered before use.

Detection on the fibre.—A solution of stannous chloride in hydrochloric acid decolourises completely. Hydrochloric acid turns the colour red; ammonia, bright yellow; concentrated sulphuric acid is coloured yellow.

Bismarck-brown, Phenylene-brown, Vesuvine, Cinnamon-brown, Canelle, Manchester-brown.

Phenylene-brown is a hydrochloride of triamidoazobenzene, possessing the formula:—

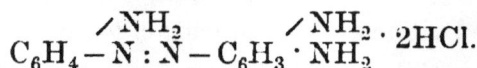

$$C_6H_4 - N : N - C_6H_3 \begin{array}{c} \diagup NH_2 \\ \diagdown NH_2 \end{array} \cdot 2HCl.$$

The colour-base is formed by the action of nitrous acid on an aqueous solution of free metaphenylenediamine, according to the equation:—

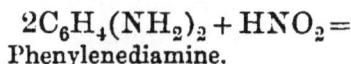

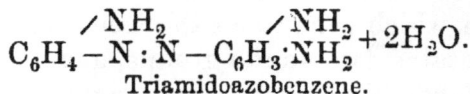

$$2C_6H_4(NH_2)_2 + HNO_2 =$$
Phenylenediamine.

$$C_6H_4 - N : N - C_6H_3 \cdot NH_2 \begin{array}{c} \diagup NH_2 \\ \diagdown NH_2 \end{array} + 2H_2O.$$
Triamidoazobenzene.

The precipitate, which still contains considerable quantities of by-products, is washed, converted into the hydrochloride, and purified.

The commercial product forms a dark powder. It contains a certain amount of by-products insoluble in water, which should be removed by filtration before use.

The free colour-base is somewhat soluble in boiling water. It can be obtained by the addition of ammonia or caustic alkali to the colour solution, in the form of a brown voluminous precipitate; when purified by recrystallisation, it forms small crystals, which melt at 137°. Although it contains three amido groups, it will not combine with more than two equivalents of acid to form stable salts. But if an excess of hydrochloric acid is added to the solution it is coloured red, owing to the formation of a salt containing three equivalents of acid. Phenylene-brown dissolves in water and alcohol with a brown colour; the solution in sulphuric acid is yellow-brown.

A mixture of stannous chloride and hydrochloric acid decolourises the aqueous solution. Basic acetate of lead produces a brown precipitate.

Application.—Phenylene-brown is used largely for dyeing leather, under various names, as vesuvine, cinnamon-brown, and Bismarck-brown. Wool is dyed without any addition, or with some Glauber's salts. The inferior qualities, which are generally used for dyeing leather, are best dissolved with the addition of a few drops of sulphuric acid. Bismarck-brown is also used for cotton which is mordanted either with tannin and tartar emetic or with Turkey-red oil. '

Detection on the Fibre.—The red colouration produced with hydrochloric acid is the best test for this colouring matter on the fibre. Ammonia produces but little change, stannous chloride decolourises.

(b) Amidoazosulphonic Acids.

The colouring matters of this class occur in commerce as alkali- or ammonia-salts. The aqueous solutions are precipitated by hydrochloric acid, in an excess of which the precipitate re-dissolves with a red colour. Ammonia and caustic soda do not produce any precipitates. Boiling alcohol does not strip the colour.

Acid-yellow, Fast-yellow, Solid-yellow, New-yellow L.

Manufacture.—Acid-yellow consists of the sodium salt of the disulphonic acid of amidoazobenzene or aniline-yellow along with some of the monosulphonic acid.

It has the composition :—

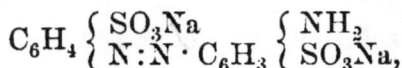

$$C_6H_4 \begin{cases} SO_3Na \\ N:N \cdot C_6H_3 \end{cases} \begin{cases} NH_2 \\ SO_3Na, \end{cases}$$

and is formed by the action of fuming sulphuric acid on amidoazobenzene.

Acid-yellow forms a yellow powder, which dissolves easily in water, but with difficulty in alcohol. On aciddulating the aqueous solution with hydrochloric acid, the free amidoazobenzene-sulphonic acid is thrown down in the form of minute needles, which re-dissolve in excess of hydrochloric acid with a reddish-yellow colour (in thin layers crimson).

Ammonia and caustic alkalies do not effect any change, nor is the colouring matter precipitated by basic acetate of lead. Sulphuric acid dissolves acid-yellow with a yellow colour. Zinc powder decolourises ; but if the colourless solution is filtered and allowed to stand in the air, the original colour is restored.

Application.—Acid-yellow is not often used for dyeing pure yellow, although it is comparatively fast as an aniline dye, and will stand steaming. The shade obtained with it is an almost pure yellow, and contains only a trace of red ; but it is not sufficiently brilliant to be used alone. On the other hand, as its colour does not incline so much towards the red, it is well adapted for dyeing compound colours, such as olive, moss-green, or brown, in the place of the natural yellow colouring matters hitherto used for this purpose. It is dyed in an acid-bath, and can therefore be combined with acid-magenta, indigo extract, fast red, etc.

Silk is dyed with boiled-off liquor, wool with sodium bisulphate, or alum and sulphuric acid. An excess of acid reddens the shade.

Large quantities of acid-yellow are used for the manufacture of diazo colours.

When dyed on the fibre, fast-yellow is reddened by hydrochloric acid; ammonia has little action; a boiling solution of stannous chloride in hydrochloric acid decolourises.

Fast-yellow R, Acid-yellow R, Yellow W, is the sodium salt of amidoazotoluenedisulphonic acid. It produces a reddish yellow on wool in an acid bath.

√*Helianthin, Gold-orange, Orange III., Methyl-orange, Tropæolin D.*

This colouring matter is the sodium salt of dimethylaniline-azobenzene-sulphonic acid, and possesses the formula—

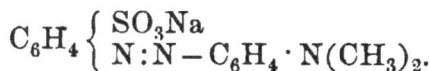

$$C_6H_4 \begin{cases} SO_3Na \\ N:N - C_6H_4 \cdot N(CH_3)_2. \end{cases}$$

It is obtained by the action of dimethylaniline on diazobenzene-sulphonic acid.

Helianthin forms an orange-yellow powder, which dissolves easily in hot water, but is only sparingly soluble in alcohol. Concentrated sulphuric acid dissolves it with a red-brown colour, which appears yellow in thin layers. With hydrochloric acid, ammonia, caustic alkali, and stannous chloride, it shows the same reactions as aniline-yellow. Basic acetate of lead throws down all the colouring matter as an orange-yellow precipitate.

If small quantities of solutions of sodium chloride or magnesium sulphate are added to a dilute solution of helianthin, the colouring matter is precipitated in the form of minute crystals.

Application.—Helianthin produces on silk and wool a fiery orange. Both these materials are dyed in an acid bath.

Helianthin, or methyl-orange, is frequently used as an indicator in volumetric analysis. The light-yellow colour

s

of the solution is immediately turned red by the addition
of a drop of hydrochloric acid.

Detection on the fibre.—Hydrochloric acid turns it red,
concentrated sulphuric acid yellow, while alkalies have no
action.

√. *Tropæolin OO, Orange IV., Orange M, Orange GS, Acid-
yellow D, New yellow, Diphenylamine orange.*

This colouring matter is formed by the action of
diazobenzene-sulphonic acid on diphenylamine:—

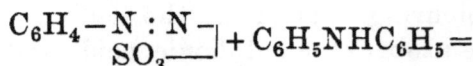

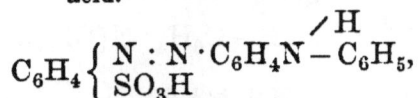

$$C_6H_4 - \underset{SO_3——}{N : N}\Big] + C_6H_5NHC_6H_5 =$$

Diazobenzene-sulphonic   Diphenylamine.
acid.

$$C_6H_4 \left\{ \begin{matrix} N : N \cdot C_6H_4N - C_6H_5, \\ SO_3H \end{matrix} \right.$$

Diphenylamine-azobenzene-sulphonic acid.

and is accordingly a phenylated acid-yellow. The commercial product is the potassium or sodium salt.

Tropæolin OO forms an orange-yellow crystalline
powder, which dissolves very easily in hot water, but is
only sparingly soluble in cold water and in alcohol. In
concentrated sulphuric acid it dissolves with a red-violet
colour. A hot aqueous solution is turned violet by hydrochloric acid, and on cooling a precipitate of the same
colour is formed. Caustic soda does not occasion any
change in the hot aqueous solution.

The colouring matter yields a very fine golden yellow
on silk or wool.

Fibres dyed with tropæolin OO are turned blue-violet
with sulphuric acid, red-violet with hydrochloric acid.

The other amidoazo dyes are described in the tables.

(c) OXYAZO DYES.

Since Peter Griess first published, in 1878, the results of his scientific investigations on the diazo and azo compounds, the manufacture of the oxyazo dyes has developed to such an extent that it forms at the present day almost the most important branch of the coal-tar colour industry. New products are constantly being brought into the market, which are as a rule superior to the older ones, either in point of brilliancy or fastness.

The immense number of colouring matters belonging to this group is easily accounted for if we consider that every primary amine belonging to the aromatic series, after having been converted into a diazo compound, will combine with almost any phenol or derivative of a phenol (in which the hydrogen atom standing opposite the hydroxyl group is not substituted) to form an azo dye.

The oxyazo dyes yield different shades of yellow, orange, red, claret, and brown. The names according to which they are known in commerce usually correspond to the colour—e.g., orange (tropæolin), scarlet, claret-red, etc. In order to distinguish between the shades, the capital letters G, GG, R, RR, etc., are used. Two samples bearing the same name and mark, but coming from different works, are, however, not always identical, and the name of a colouring matter is therefore not sufficient to prove its chemical composition. Besides these names, which are derived from the colour of the dye, there are many fancy names in use.

The description of the azo dyes given in the following pages is not a complete one, but contains all the more important products.

Manufacture of the Oxyazo Dyes.

The general method according to which all these colouring matters are prepared will be amply explained if we take xylidine-red (scarlet R) as an example. It should be mentioned here that the proportions of the different substances used correspond to the molecular weights shown in the equations.

For the preparation of about 20 kilos. of colouring matter 5 kilos. of xylidine are dissolved in 300 litres of water and twice as much hydrochloric acid as would be necessary to transform the xylidine into its hydrochloride· An aqueous solution of the calculated quantity of sodium nitrite is then gradually added, the liquid being kept in constant agitation, in order to mix the liquids as rapidly as possible. In many works the liquid is cooled with ice, in order to prevent decomposition and the formation of resinous products. In this manner a neutral solution of diazoxylene chloride is obtained, according to the equation :—

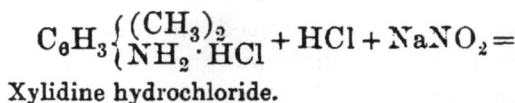

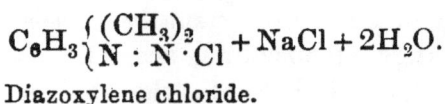

$$C_6H_3\left\{\begin{matrix}(CH_3)_2\\NH_2\cdot HCl\end{matrix}\right. + HCl + NaNO_2 =$$

Xylidine hydrochloride.

$$C_6H_3\left\{\begin{matrix}(CH_3)_2\\N:N\cdot Cl\end{matrix}\right. + NaCl + 2H_2O.$$

Diazoxylene chloride.

The liquid is now allowed to flow into another vessel of double the capacity, which contains beta-naphthol-disulphonate of soda and ammonia dissolved in 300 litres of water. Both liquids are mixed as rapidly as possible during the operation. The colouring matter is formed according to the equation :—

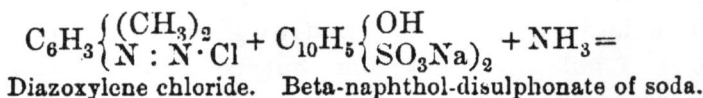

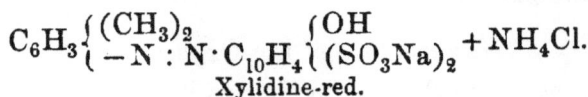

$$C_6H_3\left\{\begin{matrix}(CH_3)_2\\N:N\cdot Cl\end{matrix}\right. + C_{10}H_5\left\{\begin{matrix}OH\\SO_3Na)_2\end{matrix}\right. + NH_3 =$$

Diazoxylene chloride.   Beta-naphthol-disulphonate of soda.

$$C_6H_3\left\{\begin{matrix}(CH_3)_2\\-N:N\cdot C_{10}H_4\end{matrix}\right.\left\{\begin{matrix}OH\\(SO_3Na)_2\end{matrix}\right. + NH_4Cl.$$

Xylidine-red.

After standing some time, the liquid is heated to the boil with steam in order to separate some resinous impurities,

after which it is filtered and the colouring matter salted
out in the usual way. The precipitate is collected in a
filter-press, pressed and ground with a small quantity of
water, in order to remove the mother-liquors, after which
it is pressed again, dried and ground.

Compounds which contain the azo group $-N : N-$
twice form a special branch of the oxyazo dyes. They
are known as tetrazo dyes, and are divided into three
classes—viz., the " secondary azo compounds," the " disazo compounds," and the tetrazo compounds proper.

Secondary azo compounds are obtained when amidoazo
compounds, such as aniline-yellow or acid-yellow, are used
in place of the simple amines. A colouring matter of this
kind would, for instance, be formed according to the
following equations :—

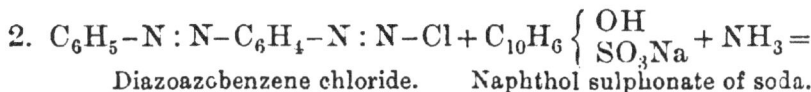

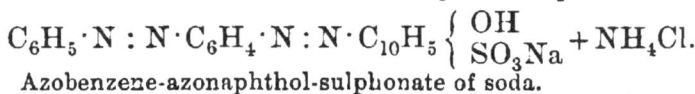

1. $C_6H_5 - N : N - C_6H_4 \cdot NH_2 \cdot HCl + HCl + NaNO_2 =$
   Amidoazobenzene hydrochloride.

   $C_6H_5 - N : N - C_6H_4 - N : N - Cl + NaCl + 2H_2O.$
   Diazoazobenzene chloride.

2. $C_6H_5 - N : N - C_6H_4 - N : N - Cl + C_{10}H_6 \begin{cases} OH \\ SO_3Na \end{cases} + NH_3 =$

   Diazoazobenzene chloride.   Naphthol sulphonate of soda.

   $C_6H_5 \cdot N : N \cdot C_6H_4 \cdot N : N \cdot C_{10}H_5 \begin{cases} OH \\ SO_3Na \end{cases} + NH_4Cl.$
   Azobenzene-azonaphthol-sulphonate of soda.

Disazo compounds.—Some of the polyatomic phenols,
like resorcin, $C_6H_4(OH)_2$, and phloroglucin, $C_6H_3(OH_3)$,
combine with more than one molecule of a diazo compound.
Thus, resorcin combines with two molecules of diazobenzene chloride to form a disazo compound having the
formula—

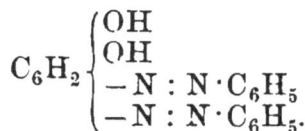

$$C_6H_2 \begin{cases} OH \\ OH \\ -N : N \cdot C_6H_5 \\ -N : N \cdot C_6H_5. \end{cases}$$

3. What are usually known at present as tetrazo dyes
are those derived from benzidine, stilbene, etc. (*see* p. 278).

Commercial name.	Scientific designation.	Diazo compound of.	Combined with.	Reaction with strong sulphuric acid.
Scarlet 4 G B. Crocëin-orange. Brilliant orange	$C_6H_5 \cdot N : N \cdot C_{10}H_5 \{^{OH}_{SO_3Na}}$ Benzeneazo-β-naphtholsulphonate of soln.	Aniline.	β-naphtholmono-sulphonic acid S.	Orange-yellow.
Orange G.	$C_6H_5 \cdot N : N \cdot C_{10}H_4 \{^{OH}_{(SO_3Na)_2}}$ Benzeneazo-β-naphtholdisulphonate of soda.	Aniline.	β-naphtholdisulphonic acid G.	Orange-yellow.
Scarlet 2 G.	$C_6H_5 \cdot N : N \cdot C_{10}H_4 \{^{OH}_{(SO_3Na)_2}}$ Benzeneazo-β-naphtholdisulphonate of soda.	Aniline.	β-naphtholdisulphonic acid R.	Cherry-red.
Acid-brown G.	$C_8H_5 \cdot N : N \cdot C_6H_2 \{^{NH_2}_{NH_2}\}\{N : N \cdot C_6H_4 \cdot SO_3Na}$ Benzene-azometadiamidoazobenzenesulphonate of soda.	Anilino.	Metadiamidoazo-benzeneparasulphonic acid.	Red-brown.
Orange III. Orange, No. 3.	$C_6H_4 \{^{NO_2}_{N : N \cdot C_{10}H_4 \{^{OH}_{(SO_3Na)_2}}}}$ Nitrobenzeneazo-β-naphtholdisulphonate of soda.	m. Nitraniline.	β-naphtholdisulphonic acid R.	Orange-yellow.
Orchil substitute (Poirrier).	$C_6H_4 \{^{NO_2}_{N : N \cdot C_{10}H_5 \{^{NH_2}_{SO_3Na}}}}$	p. Nitraniline.	Naphthionic acid.	Magenta.

$C_6H_3\left\{{(CH_3)_2 \atop N:N\cdot C_{10}H_5\left\{{OII \atop SO_3Na}\right.}\right.$ Xyleneazo-α-naphtholsulphonate of soda.	Xylidine.	α-naphtholmonosulphonic acid N.W.	Magenta.
$C_6H_3\left\{{(CH_3)_2 \atop N:N\cdot C_{10}H_4\left\{{OH \atop (SO_3Na)_2}\right.}\right.$ Xyleneazo-α-naphtholdisulphonate of soda.	Xylidine.	α-naphtholdisulphonic acid (Schöllkopf Aniline Co.).	Cherry-red.
$C_6H_3\left\{{(CH_3)_2 \atop N:N\cdot C_{10}H_5\left\{{OH \atop SO_3Na}\right.}\right.$ Xyleneazo-β-naphtholsulphonate of soda.	Xylidine.	β-naphtholmonosulphonic acid S.	
$C_6H_3\left\{{(CH_3)_2 \atop N:N\cdot C_{10}H_4\left\{{OH \atop (SO_3Na)_2}\right.}\right.$ Xyleneazo-β-naphtholdisulphonate of soda.	Xylidine. (impure).	β-naphtholdisulphonic acid R.	

Commercial name.	Scientific designation.	Diazo compound of.	Combined with.	Reaction with strong sulphuric acid.
Scarlet 2 R. *Xylidine-red.* *Xylidine-scarlet.*	$C_6H_3\begin{Bmatrix}(CH_3)_2 \\ N:N\cdot C_{10}H_4\begin{Bmatrix}OH \\ (SO_3Na)_2\end{Bmatrix}\end{Bmatrix}$ Xyleneazo-β-naphtholdisulphonate of soda.	Xylidine. (From crystallised xylidine hydrochloride).	β-naphtholdisulphonic acid R.	Cherry-red.
Resorcin-brown.	$C_6H_3\begin{Bmatrix}(CH_3)_2 \\ N:N\cdot C_6H_2\begin{Bmatrix}OH \\ OH \\ N:N\cdot C_6H_4\cdot SO_3Na\end{Bmatrix}\end{Bmatrix}$ Xylenazoresorcinazosulphanilate of soda.	Xylidine.	Tropæolin O.	Brown.
Scarlet 3 R. *Cumidine-red.* *Cumidine-scarlet*	$C_6H_2\begin{Bmatrix}(CH_3)_3 \\ N:N\cdot C_{10}H_4\begin{Bmatrix}OH \\ (SO_3Na)_2\end{Bmatrix}\end{Bmatrix}$ Cumenazo-β-naphtholdisulphonate of soda.	ψ Cumidine.	β-naphtholdisulphonic acid R.	Cherry-red.
Scarlet 3 R.	$C_6H_2\begin{Bmatrix}(CH_3)_2 \\ C_2H_5 \\ N:N\cdot C_{10}H_4\begin{Bmatrix}OH \\ (SO_3Na)_2\end{Bmatrix}\end{Bmatrix}$ Ethyldimethylbenzeneazo-β-naphtholdisulphonate of soda.	Amidoethyldimethylbenzene.	β-naphtholdisulphonic acid R.	Cherry-red.

Dye	Formula	Base	Acid	Colour
Crystal-scarlet 6 R. New coccin R.	$C_{10}H_7 \cdot N : N - C_{10}H_4 \{ \begin{smallmatrix} OH \\ (SO_3Na)_2 \end{smallmatrix}$ a-naphthaleneazo-β-naphtholdisulphonate of soda.	α-naphthylamine	β-naphtholdisulphonic acid G.	Violet.
Fast-red B. Claret-red B.	$C_{10}H_7 \cdot N : N \cdot C_{10}H_4 \{ \begin{smallmatrix} OH \\ (SO_3Na)_2 \end{smallmatrix}$ a-naphthaleneazobetanaphtholdisulphonate of soda.	α-naphthylamine	β-naphtholdisulphonic acid R.	Blue.
Carmin-naphtha.	$C_{10}H_7 \cdot N : N \cdot C_{10}H_6OH$ β-naphthaleneazo-β-naphthol.	β-naphthylamine.	β-naphthol.	Magenta.
Brilliant croceïn.	$C_6H_5 \cdot N : N \cdot C_6H_4 \cdot N : N \cdot C_{10}H_4 \{ \begin{smallmatrix} OH \\ (SO_3Na)_2 \end{smallmatrix}$ Benzeneazobenzeneazo-β-naphtholdisulphonate of soda.	Amidoazobenzene.	β-naphtholdisulphonic acid γ.	Red-Violet.
Cloth-red G.	$C_6H_4 \{ \begin{smallmatrix} CH_3 \\ N : N \cdot C_6H_3 \end{smallmatrix} \{ \begin{smallmatrix} CH_3 \\ N : N \cdot C_{10}H_5 \end{smallmatrix} \{ \begin{smallmatrix} OH \\ SO_3Na \end{smallmatrix}$ Tolueneazotolueneazo-β-naphtholsulphonate of soda.	Amidoazotoluene.	β-naphtholmonosulphonic acid S.	Blue.
Archil-red A.	$C_6H_6 \{ \begin{smallmatrix} (CH_3)_2 \\ N : N \cdot C_6H_2 \end{smallmatrix} \{ \begin{smallmatrix} (CH_3)_2 \\ N : N \cdot C_{10}H_4 \end{smallmatrix} \{ \begin{smallmatrix} OH \\ (SO_3Na)_2 \end{smallmatrix}$ Xyleneazoxyleneazo-β-naphtholdisulphonate of soda	Amidoazoxylene.	β-naphtholdisulphonic acid R.	Dark-blue.

Commercial name.	Scientific designation.	Diazo compound of.	Combined with.	Reaction with strong sulphuric acid.
Resorcin-yellow. Tropæolin O. Tropæolin R. Chrysoïn, Chryseolin. Yellow T. Golden-Yellow.	$C_6H_4\begin{cases}SO_3Na\\N:N\cdot C_6H_3\end{cases}\begin{cases}OH\\OH\end{cases}$ Benzenesulphonate of soda azoresorcin.	Sulphanilic acid.	Resorcin.	Yellow.
Orange I. α - Naphthol-orange. Tropæolin 000 No. 1.	$C_6H_4\begin{cases}SO_3Na\\N:N\cdot C_{10}H_6OH\end{cases}$ Benzenesulphonate of soda-azo-α-naphthol.	Sulphanilic acid.	α-naphthol.	Magenta.
Orange II. β - Naphthol-orange Tropæolin 000 No. 2. Mandarin. Chrysaurein. Gold-orange.	$C_6H_4\begin{cases}SO_3Na\\N:N\cdot C_{10}H_6OH\end{cases}$ Benzenesulphonate of soda-azo-β-naphthol.	Sulphanilic acid.	β-naphthol.	Magenta.
Orange III. Methyl-orange. Dimethylaniline Orange Tropæolin D.	$C_6H_4\begin{cases}SO_3Na\\N:N\cdot C_6H_4N(CH_3)_2\end{cases}$ Benzenesulphonate of soda azodimethylaniline.	Sulphanilic acid.	Dimethylaniline.	Brown.

Dyes	Constitution			Colour
Diphenylamine orange. Orange IV. Tropæolin OO. Orange M. Acid-yellow. Orange G. S. New yellow.	$C_6H_4\{N:N \cdot C_6H_4 \cdot NH \cdot C_6H_5$ Benzenesulphonate of soda azodiphenylamine.			
Metanil-yellow. Orange M. N.	$C_6H_4\{\begin{array}{l}SO_3Na\\N:N \cdot C_6H_4 \cdot NH \cdot C_6H_5\end{array}$ Meta-benzenesulphonate of soda azodiphenylamine.	Meta-amidobenzenesulphonate of soda.	Diphenylamine.	Violet.
Orange T. Mandarin G R. Orange R.	$C_6H_3\{\begin{array}{l}SO_3Na\\CH_3\\N:N \cdot C_{10}H_6OH\end{array}$ Orthotoluenesulphonate of soda-azo-β-naphthol.	Orthotoluidine-sulphonic acid.	β-naphthol.	Magenta.
Fast-red A. Roccellin. Rubidin.	$C_{10}H_6\{\begin{array}{l}SO_3Na\\N:N \cdot C_{10}H_6OH\end{array}$ Naphthalenesulphonate of soda azo-β-naphthol.	Naphthionic acid.	β-naphthol.	Violet.
Azorubine S. Fast-red C. Carmoisin.	$C_{10}H_6\{N:N \cdot C_{10}H_5\{\begin{array}{l}OH\\SO_3Na\end{array}$ Naphthalenesulphonate of soda-azo-α-naphthol-sulphonate of soda.	Naphthionic acid.	α naphtholsulphonic acid N.W.	Violet.
Crocein 3 B X.	$C_{10}H_6\{N:N \cdot C_{10}H_5\{\begin{array}{l}OH\\SO_3Na\end{array}$ Naphthalenesulphonate of soda azo-β-naphtholsulphonate of soda.	Naphthionic acid.	β-naphtholsulphonic acid B.	Red-violet.

Commercial name.	Scientific designation.	Diazo compound of.	Combined with.	Reaction with strong sulphuric acid.
Fast-red E.	$C_{10}H_6\{^{SO_3Na}_{N:N \cdot C_{10}H_5}\{^{OH}_{SO_3Na}$	Naphthionic acid.	β-naphtholsulphonic acid.	
New coccin Brilliant-Scarlet. Cochineal red A.	$C_{10}H_6\{^{SO_3Na}_{N:N \cdot C_{10}H_4}\{^{OH}_{(SO_3Na)_2}$ Naphthalenesulphonate of soda azobetanaphthol-disulphonate of soda.	Naphthionic acid.	β-naphtholdisulphonic acid G.	Magenta.
Fast-red D. Claret-red S. Amaranth.	$C_{10}H_6\{^{SO_3Na}_{N:N \cdot C_{10}H_4}\{^{OH}_{(SO_3Na)_2}$	Naphthionic acid.	β-naphtholdisulphonic acid R.	Violet.
Scarlet 6 R.	$C_{10}H_6\{^{SO_3Na}_{N:N \cdot C_{10}H_3}\{^{OH}_{(SO_3Na)_3}$	Naphthionic acid.	β-naphtholtrisulphonic acid.	Violet.

Crocëin-scarlet 7 B. Scarlet 6 R. B.	$C_6H_3 \begin{cases} CH_3 \\ SO_3Na \\ N:N \cdot C_6H_3 \begin{cases} CH_3 \\ N:N \cdot C_{10}H_5 \begin{cases} OH \\ SO_3Na \end{cases} \end{cases} \end{cases}$ Toluenesulphonate of soda azo-toluenesulphonate of soda-azo-β-naphtholsulphonate of soda.	Amidoazotoluene-sulphonic acid.	β-naphtholsul-phonic acid B.	Blue.
Blue-black B. Azo-black.	$C_{10}H_6 \begin{cases} SO_3Na \\ N:N \cdot C_{10}H_6 \cdot N:N \cdot C_{10}H_4 \begin{cases} OH \\ (SO_3Na)_2 \end{cases} \end{cases}$ Naphthalenesulphonate of soda azonaphthaleneazo-β-naphtholdisulphonate of soda.	Sulphonic acid of β-naphtha-lenazo-α-naph-thylamine.	β-naphtholdi-sulphonic acid R.	Blue-green.
Naphthol-black.	$C_{10}H_5 \begin{cases} (SO_3Na)_2 \\ N:N \cdot C_{10}H_6 \cdot N:N \cdot C_{10}H_4 \begin{cases} OH \\ (SO_3Na)_2 \end{cases} \end{cases}$ Naphthalenedisulphonate of soda azonaphthalene-azo-β-naphtholdisulphonate of soda.	Amidoazonaph-thalenedisul-phonic acid.	β-naphtholdi-sulphonic acid R.	Green.

The tables on pp. 262–269 contain all the most important oxyazo dyes which are or have been used on a large scale in practice.

The first column contains the commercial names of the colouring matters, while in the second column their chemical constitution and scientific denomination is given. In the third column are the amido compounds, which are first diazotised in the ordinary manner, and are then combined with the phenols in the fourth column, producing the respective dyes. The last column gives the colour reaction with strong sulphuric acid which is characteristic, and frequently serves for the detection of the dye either as such or on the fibre.

Properties of the Oxyazo Dyes.

The commercial products are yellow, orange, scarlet or brown powders, which are either crystalline in themselves or can generally be crystallised from water or alcohol, or a mixture of the two. They are nearly all soluble in water, and most of them will also dissolve in alcohol.

Their behaviour with concentrated sulphuric acid, in which all azo dyes dissolve with very intense colours, is especially characteristic. The colours of these solutions differ so widely that they form for a practised eye one of the most valuable means of distinguishing the colouring matters. Thus, the scarlets R, RR, 3R, S and SS dissolve in sulphuric acid with an *orange* colour, which appears *crimson* in thin layers.

Scarlet G and tropæolin O yield *yellow* or *orange-yellow* solutions.

Biebrich-scarlet colours the acid *green ;*

Crocëin-scarlet, *blue ;* and

Tropæolin 3O, or fast-red, both yield a *violet.*

The reaction with concentrated sulphuric acid further enables us in many cases to obtain an idea of the con-

stitution of a colouring matter. Thus the following
rule has been adopted for all tetrazo dyes which are
derived from beta-naphthol-azobenzene-azobenzene—

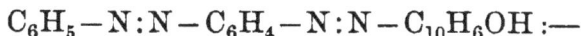

$$C_6H_5 - N:N - C_6H_4 - N:N - C_{10}H_6OH :—$$

" Colouring matters which contain the sulpho group
in the benzene ring only dissolve in concentrated
sulphuric acid with a green colour (Biebrich-scarlet).
" If the sulpho group is only contained in the naphthol
the solution in sulphuric acid will be red or violet
(Scarlet S).
" When, however, sulpho groups are contained both in
the benzene ring and in the naphthol, they are
coloured blue with sulphuric acid."

The aqueous solutions of the oxyazo dyes are either
not changed at all when acidulated with hydrochloric
acid, or a precipitate is formed. The latter is especially
the case when the colouring matter only contains one
sulpho group. The formation of a precipitate is due to
the fact that the free acid is insoluble in water, as in the
case of alkali-blue. The free acids of those colouring
matters, however, which contain two or more sulpho
groups are also soluble in the free state, and are therefore not precipitated by acids. Tropæolin 3O gives a
purple precipitate, which dissolves in excess of hydrochloric acid. Some of the Biebrich-scarlets show the
same reaction.

Ammonia and caustic alkalies do not, as a rule, cause any
precipitates in dilute solutions; but they often change
the colour, owing to a saturation of the free hydroxyl
groups.

A dilute aqueous and only very light yellow solution
of tropæolin 3O No. 2 shows this change of colour very
readily; the slightest trace of alkali suffices to turn the
colour of the liquid to a fine crimson. For this reason

the product is useful as an indicator in volumetric analysis.

Scarlets G and R, crocëin-scarlet and Biebrich-scarlet show a similar reaction, but are not nearly so sensitive as tropæolin 3O No. 2. Ammonia is almost without action on the solutions of scarlet 2R and scarlet 3R.

Concentrated solutions of the oxyazo dyes are precipitated by metallic salts; most of the precipitates obtained in this manner (as, for instance, those with alumina and lead) are soluble in excess of water; consequently these colouring matters are not adapted for the preparation of lakes.

By energetic reducing agents, the oxyazo dyes are split up like the amidoazo dyes, two or more amido compounds being formed. Thus, if Biebrich-scarlet is treated with stannous chloride and hydrochloric acid, it yields sulphanilic acid, phenylene diamine, and beta-amidonaphthol, according to the equation:—

$$C_6H_4 \overset{\diagup SO_3H}{-} N:N - C_6H_3 \overset{\diagup SO_3H}{-} N:N - C_{10}H_6OH + 4H_2 =$$

Beta-naphthol-azobenzene—sulphonic acid—azobenzene-sulphonic acid.

$$C_6H_4\overset{\diagup SO_3H}{NH_2} | + C_6H_4(NH_2)_2 + C_{10}H_5 \begin{cases} OH \\ SO_3H \\ NH_2. \end{cases}$$

Sulphanilic acid.   Phenylene diamine.   Amidonaphthol-sulphonic acid.

This reaction is of great value in the scientific examination of an azo dye for ascertaining the chemical composition; but the separation, purification, and examination of the decomposition products takes up so much time, and is so difficult, that the work can only be carried out by a competent chemist. The complete decolourisation by means of stannous chloride and hydrochloric acid can, however, be used for distinguishing the azo dyes from other yellow and red dyes, especially the natural ones.

Alkaline reducing agents decompose most of the oxyazo dyes in the cold. In some cases, however, especially with some of the tetrazo dyes, their action is not as energetic as that of stannous chloride.

The tetrazo dyes derived from the azobenzene-sulphonic acids (acid-yellow), such as Biebrich-scarlet and crocëin-scarlet, can easily be recognised by their behaviour towards zinc powder, which rapidly decolourises the alkaline solution. If some of the decolourised liquid is poured on to a watch-glass (in order to present a large surface to the action of the air), it becomes intensely yellow, especially if diluted with a little water. If the liquid is filtered and is then acidulated with hydrochloric acid, a yellow crystalline precipitate is formed, which dissolves in excess of hydrochloric acid with a red colour, thus showing the reactions of acid-yellow.

The reduction with zinc powder in alkaline solution, therefore, only effects a partial decomposition. Taking Biebrich-scarlet as an example, the following equation would be fulfilled :—

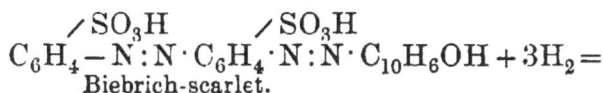

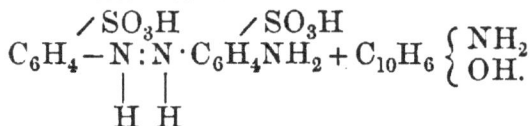

$$C_6H_4 \underset{\text{Biebrich-scarlet.}}{- N : N \cdot C_6H_4 \cdot N : N \cdot C_{10}H_6OH} + 3H_2 =$$

with SO_3H groups shown on C_6H_4.

$$C_6H_4 - \underset{\underset{H}{|}}{N} : \underset{\underset{H}{|}}{N} \cdot C_6H_4NH_2 + C_{10}H_6 \begin{cases} NH_2 \\ OH. \end{cases}$$

Amidohydrazobenzene-disulphonic   Amidonaphthol.
acid.

The colourless hydrazo compound is oxidised in the air to the corresponding azo compound (fast-yellow).

Application of the Oxyazo Dyes.

Among the tropæolins, tropæolin R and tropæolin No. 2 are used to a considerable extent for dyeing silk orange. The scarlets have replaced cochineal to a considerable extent in wool-dyeing, while some (crocëin-scarlet, etc.) are used very largely in cotton-dyeing.

T

As a rule the fastness of the azo dyes increases with the molecular weight. Most of them are pretty fast to light, but they cannot be used for dyeing the Turkish fez, the colour of which must be so fast that several months' exposure to the direct sunlight will not bleach it.

Thus xylidine-scarlet (scarlet R) when exposed for eight months to direct sunlight is changed to a light and dull pink; but it stands the action of diffused sunlight very well. The azo dyes do not stand washing with soap as well as cochineal, nor are they as fast to milling as the latter. This is explained by the fact that cochineal-scarlet is an alumina-tin lake, insoluble in water, while the compounds of the azo dyes with alumina and tin dissolve in water.

Silk is dyed in a bath containing boiled-off liquor, or in a soap-bath acidulated with sulphuric acid.

Wool is dyed in an acid bath. The colouring matters dye more evenly and are more completely extracted from the solution when certain salts are added to the dye-bath, in order to decrease their solubility. Glauber's salts and sulphuric acid, or sodium bisulphate and acetic acid, are generally used for this purpose. Some of the azo dyes are used with alum or stannic chloride and tartar; the shades obtained in this matter are purer and more brilliant than when dyed according to the ordinary method.

Cotton can be prepared with an oil mordant or with stannate of soda and alum. It is then dyed in a concentrated solution of the colouring matter, and dried without being washed. The Biebrich- and crocëin-scarlets can also be dyed on unprepared cotton by padding or otherwise treating the fabric with a concentrated solution of the colouring matter and alum at a moderate temperature. The colours do not stand washing.

For cotton-warp dyeing and in calico printing, Holliday's and Graessler's patents are of importance; in both processes the formation of the colouring matter is effected directly on the fibre. According to Holliday's method cotton may be dyed orange, scarlet, maroon, etc., according to the amido compound and phenol used. The material is first prepared with Turkey-red oil, then passed through an alkaline solution of the phenol (*e.g.*, beta-naphthol), after which it is squeezed and passed through a solution of a diazo compound prepared in the ordinary way (*e.g.*, diazoxylene chloride), when the colouring matter is precipitated directly on the fibre. As no sulpho compounds are used in this process, the colouring matters precipitated on the fibre are as a rule very fast, owing to their insolubility in water, dilute acids, and alkalies. The scarlets produced in this manner are as brilliant as Turkey-red, but are not quite as fast as the latter, especially to heat, which destroys or volatilises the colour. The goods cannot therefore undergo the operation of " hot pressing."

The following example will serve to illustrate Graessler's patent. In order to produce xylidine-red, $C_6H_3(CH_3)_2$ $-N:N-C_{10}H_6OH$, on the fibre a mixture of beta-naphthol, xylidine, sodium nitrite and ammonium chloride, thickened with starch-paste, is printed and steamed. The heat causes the xylidine to decompose the ammonium chloride, when ammonia is liberated and xylidine hydrochloride is formed, which latter reacts with the sodium nitrite so as to form sodium chloride and diazoxylene chloride—

$$C_6H_3 \begin{cases} (CH_3)_2 \\ -N:\overset{2}{N}\cdot OH. \end{cases}$$

This combines in the nascent state with the naphthol to form xylidine-red. The whole reaction is expressed by the following equations :—

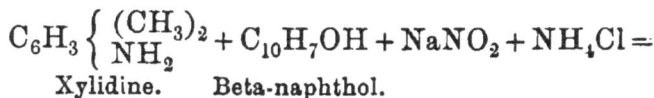

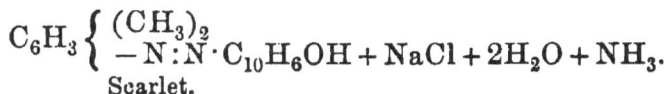

$$C_6H_3 \begin{cases} (CH_3)_2 \\ NH_2 \end{cases} + C_{10}H_7OH + NaNO_2 + NH_4Cl =$$
Xylidine.   Beta-naphthol.

$$C_6H_3 \begin{cases} (CH_3)_2 \\ -N:N\cdot C_{10}H_6OH + NaCl + 2H_2O + NH_3. \end{cases}$$
Scarlet.

According to Dawson's patent, the azo colours may be fixed on the fibre in one operation. For this purpose an aromatic amine (for example, naphthylamine) is diazotised in the usual manner, after which it is completely neutralised by the addition of chalk. To the solution obtained in this manner the equivalent of naphthol is added in a very finely divided state, along with acetate of soda and a slight quantity of acetic acid. Cotton immersed in such a bath acts catalytically, and induces the gradual formation of the colour upon its surface or within the fibre, thereby becoming gradually and effectively dyed. Wool and silk may be similarly treated.

Detection on the Fibre.

When dyed on the fibre, all orange, red, claret-red, and brown oxyazo dyes are completely and permanently decolourised * when the material is boiled with tin or zinc powder and hydrochloric acid. When treated with concentrated sulphuric acid they yield characteristic intense colourations, which have already been described in the tables.

Bisulphite Compounds of the Azo Dyes.

Several patents have been taken out for the production of soluble compounds from the insoluble azo dyes by treatment with alkaline bisulphites. The preparation of these compounds is either effected like that of coeruleïn S, or alizarin-blue S, by treating the finely divided colouring matter with a concentrated solution of an alkaline bisulphite, or by treating the azo dye with a bisulphite in a solvent common to both (alcohol). The compounds are not affected by dilute acids, but under the influence of heat they are decomposed into the original dye-stuff and a neutral sulphite. It has been proposed to utilise this

---

* In one or two cases the liquid, after filtering, is turned yellow in the air.

latter property in calico-printing, for which purpose the bisulphite compound, suitably thickened, is printed on the fabric and subsequently steamed. During the steaming the decomposition takes place, and the insoluble azo dye is fixed on the fibre. The colours produced in this manner are said to stand washing.

The most important commercial product of this class which has hitherto been brought out is known as Azarin.

Azarin possesses the formula—

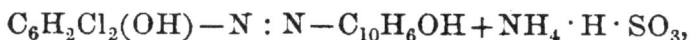

$$C_6H_2Cl_2(OH) - N : N - C_{10}H_6OH + NH_4 \cdot H \cdot SO_3,$$

by which it is represented as a molecular compound of naphthol-azo-dichlor-phenol and ammonium bisulphite. It forms an orange-paste, which resembles ordinary alizarin-paste in appearance. The colouring matter is not easily soluble in water, but dissolves readily in alkalies with a deep bluish-violet colour. If the paste is heated, a copious disengagement of sulphur dioxide takes place and the colour turns to a scarlet.

Azarin is chiefly applied to cotton. The material is first prepared with Turkey-red oil and aluminium acetate, after which it is dyed in a neutral bath of azarin. A brilliant red inclining to a crimson can be obtained in this manner, which stands the action of soap tolerably well.

The same principle has recently been applied (and patented) for rendering soluble the colours produced from diazotised primuline (*see* p. 289, and the phenols. The great advantage possessed by the dyes obtained in this manner is that they may be applied in one bath.

Azo Dyes derived from Benzidine, Tolidine, and Stilbene.

A. *Colours derived from Benzidine and Tolidine.*

If benzidine,—

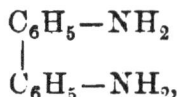

$$C_6H_5 - NH_2$$
$$|$$
$$C_6H_5 - NH_2,$$

is treated with nitrous acid, it yields a tetrazo compound, the hydrochloride of which would be represented by the formula—

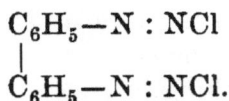

$$C_6H_5—N : NCl$$
$$|$$
$$C_6H_5—N : NCl.$$

When combined in the ordinary way with the aromatic amines or phenols, or their derivatives, the salts of tetrazo-diphenyl yield a series of colouring matters which possess the peculiar but valuable property of dyeing vegetable fibres in a neutral or alkaline bath without the intervention of a mordant.

✓ *Congo-red* is obtained according to the patent specification by the action of tetrazo-diphenyl chloride on naphthylamine-sulphonate of soda, and possesses the formula—

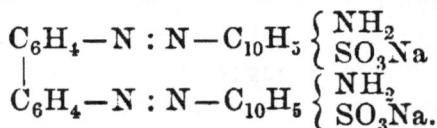

$$C_6H_4—N : N—C_{10}H_5 \begin{cases} NH_2 \\ SO_3Na \end{cases}$$
$$|$$
$$C_6H_4—N : N—C_{10}H_5 \begin{cases} NH_2 \\ SO_3Na. \end{cases}$$

The commercial product is a brown-red powder, which dissolves easily in water with a fine red colour. The aqueous solution is so sensitive to acids that a single drop of very dilute sulphuric acid suffices to convert the whole of the liquid to a beautiful blue; in strong solutions a precipitate of the free sulphonic acid is formed. This property makes Congo-red valuable as an indicator in volumetric analysis.

Application.—Cotton is either dyed in a neutral bath or in a weak soap-bath at a boiling temperature. A fine scarlet is produced in this manner; but although fast to soap, it possesses the fatal property of being changed by the slightest trace of acid to a violet or even to a blue. Washing restores the original shade. Wool can be dyed in a neutral bath. Another method is to dye in an acid bath, and subsequently brighten the colour in a weak alkali.

Brilliant Congo G is formed by the action of one molecule each of β-naphthylamine disulphonic acid and β-naphthylamine monosulphonic acid on tetrazobenzene chloride. *Brilliant Congo* R is formed in a similar manner, tolidine being substituted for benzidine. *Congo* 4 R is the product of the reaction of equal molecules of naphthionic acid and resorcin on one molecule of diazotised tolidine.

Benzopurpurin 4 B is formed from diazotised tolidine (1 mol.) and naphthionic acid (2 mol.). *Benzopurpurin* 6 B is the product of the action of Laurent's α-naphthylamine monosulphonic acid (2 mol.) on diazotised tolidine (1 mol.). *Deltapurpurin* 7 B is formed from diazotised tolidine (1 mol.) and β-naphthylamine monosulphonic acid S (2 mol.). *Rosazurin* B is formed by the action of methyl-β-naphthylamine sulphonic acid (2 mol.) on diazotised tolidine (1 mol.).

All these colouring matters are applied like Benzopurpurin, and produce various shades of red. Phosphate of soda, stannate of soda, or Turkey-red oil, are sometimes added to the dye-bath in order to obtain more brilliant colours. The dye-bath should be as concentrated as possible.

Chrysamin is a yellow dye obtained by the action of tetrazo-diphenyl chloride on salicylate of soda. It possesses the formula—

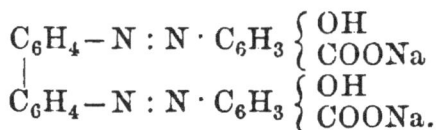

$$C_6H_4 - N : N \cdot C_6H_3 \left\{ \begin{array}{l} OH \\ COONa \end{array} \right.$$
$$C_6H_4 - N : N \cdot C_6H_3 \left\{ \begin{array}{l} OH \\ COONa. \end{array} \right.$$

Chrysamin forms a yellow powder, which is sparingly soluble in cold water, but dissolves easily on boiling, with an orange colour. An addition of caustic soda turns the colour of the solution to an orange-red, from which sulphuric acid precipitates the colouring matter (the free acid) in orange flakes. Concentrated sulphuric acid dissolves chrysamin with a deep magenta colour.

Application.—Cotton is dyed with the addition of soap and phosphate of soda at a boiling temperature. The shades obtained are fast to light.

Chrysamin R is the corresponding compound from tolidine.

Benzopurpurin is obtained by the action of tetrazo-ditolyl chloride on naphthylamine sulphonate of soda, and possesses the formula—

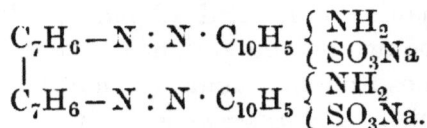

$$C_7H_6 - N : N \cdot C_{10}H_5 \begin{cases} NH_2 \\ SO_3Na \end{cases}$$
$$C_7H_6 - N : N \cdot C_{10}H_5 \begin{cases} NH_2 \\ SO_3Na. \end{cases}$$

Benzopurpurin forms a dark-red powder, which dissolves easily in water with a red-orange colour, on which caustic soda is without action. Dilute sulphuric acid precipitates the colouring matter from its solutions as a brown-red precipitate, resembling ferric hydrate in appearance. Concentrated sulphuric acid dissolves the colouring matter with a pure blue colour.

Application. Cotton is dyed with the addition of potassium carbonate at a boiling temperature. A very fine scarlet may be obtained in this manner, which differs materially from the red obtained with Congo-red in not being changed by dilute acids and in being faster to light.

Azo-blue is formed by the action of tetrazo-ditolyl chloride on beta-naphthol sulphonate of soda, and possesses the formula—

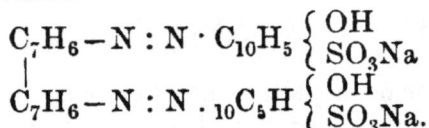

$$C_7H_6 - N : N \cdot C_{10}H_5 \begin{cases} OH \\ SO_3Na \end{cases}$$
$$C_7H_6 - N : N \cdot {}_{10}C_5H \begin{cases} OH \\ SO_3Na. \end{cases}$$

Azo-blue forms a dark-blue powder, which dissolves easily in water, with a rich violet colour. Caustic soda turns the colour of the solution to a fine crimson, which is restored to the original colour by the addition of dilute sulphuric acid.

Concentrated sulphuric acid dissolves azo-blue with a pure blue colour.

Application.—Cotton is dyed with the addition of soap and phosphate of soda at a boiling temperature. A reddish blue is thus obtained, which is fast to soap and acids, but is not very brilliant.

Azo-violet is the product of the reaction of equal molecules of naphthionic acid and a-naphthol monosulphonic acid NW on one molecule diazotised dianisidine. *Heliotrop* is produced from methyl-β-naphthylamine sulphonic acid S (2 mol.) and diazotised anisidine (2 mol). *Benzazurin* G is the purest shade of blue hitherto produced in this series. It is formed from a-naphthol monosulphonic acid NW (2 mol.) and diazotised dianisidine (1 mol.). The application of these colouring matters in dyeing resembles that of azo-blue.

The colouring matters of this group may be used along with each other for the production of mixed shades on cotton. They are also applicable to wool and silk under the same conditions, and may therefore be found useful in the dyeing of mixed fabrics.

B. *Colours derived from Stilbene.*

By diazotising diamidostilbene disulphonic acid,—

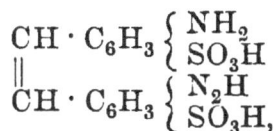

$$CH \cdot C_6H_3 \left\{ \begin{array}{l} NH_2 \\ SO_3H \end{array} \right.$$
$$\| \qquad \qquad \qquad$$
$$CH \cdot C_6H_3 \left\{ \begin{array}{l} N_2H \\ SO_3H, \end{array} \right.$$

and combining the product formed with various phenols, or aromatic amines, a series of colours is obtained which resemble in their properties those derived from benzidine and tolidine.

The diazotised diamidostilbene disulphonic acid (1 mol.) yields with—

β-naphthylamine (2 mol.), *Hessian-purple* N ; with naphthionic acid (2 mol.), *Hessian-purple* P ; with β-naphthylamine monosulphonic acid Br, *Hessian-purple* B ; and with β-naphthylamine monosulphonic acid D, *Hessian-purple* D. All these dyes yield blue shades of red, and are applied on cotton either with soap or with salt and acetic acid.

With salicylic acid (2 mol.) the diazotised diamido-stilbene disulphonic (1 mol.) acid yields *Hessian-yellow,* and with phenol (2 mol.) *Brilliant-yellow* is obtained. By ethylating brilliant-yellow *Chrysophenin* is formed, which will probably, by reason of its superiority, take the place of the older dye Chrysamin.

Hessian-violet is formed from diazotised diamidostilbene disulphonic acid (1 mol.) and equal molecules of α-naphthylamine and β-naphthol.

Mikado-orange is formed by the action of glycerin in alkaline solution on paranitrotoluene sulphonic acid. Its exact composition is not yet known, but it probably also belongs to the stilbene colours. The other so-called *mikado colours* are obtained in a similar way, other oxidisable substances being used in place of glycerin.

Primuline.

This colouring matter, which was first brought out by Messrs. Brooke, Simpson & Spiller, is formed by the action of sulphur on paratoluidine and subsequent sulphonation. It is the monosulphonic acid of the base, $C_{28}H_{18}N_4S_3$.

The commercial product forms a lemon-yellow powder, soluble in water with a yellow colour; in strong sulphuric acid the solution has a yellow colour and a strong green fluorescence.

Application.—If cotton, wool, or silk are boiled in a solution of primuline, a canary-yellow is obtained, which is in itself not of much interest. If, however, the material dyed with primuline is passed into a bath containing nitrous acid (nitrite of soda and sulphuric acid), and then into an alkaline solution of a phenol,* various shades are produced on the fibre which range from yellow to red and brown, and which are perfectly fast to acids and alkalies, but not to light. As is the case with the benzidine colours, primuline is more readily absorbed by the vegetable fibres from an alkaline bath, while the presence of a weak acid favours its absorption by the animal fibres. In dyeing with primuline, it is necessary to preserve the diazotised fibre from the action of light, since an exposure of less than half a minute to direct sunlight is sufficient to destroy the diazo compound, and thus to produce uneven results.

Detection.—Treated with an alkaline solution of hyposulphite of soda ($NaSO_2$), all the primuline colours are converted back again into the original canary-yellow shade, which, when diazotised and passed into an alkaline solution of β-naphthol, is turned scarlet.

Primuline also comes into the market as *Sulphine, Polychromine,* and *Carnotine.*

DERIVATIVES OF QUINOLINE AND ACRIDINE.

Flavaniline.

If acetanilid, $C_6H_5NHC_2H_3O$, is heated for several hours with zinc chloride to 250°–270°, a brown melt is obtained, which yields when purified a yellow powder, soluble in water. The free base has the composition $C_{16}H_{14}N_2$, and melts at 97°. The combination, with one equivalent of hydrochloric acid, forms the colouring matter flavaniline a-para-amido phenyl-γ-lepidine hydrochloride, $C_{16}H_{14}N_2$ HCl, or—

$$C_6H_4 \begin{cases} \overset{\displaystyle CH_3}{\underset{\displaystyle |}{C}} = CH \\ | \\ N = C \cdot C_6H_4 \cdot NH_2 \cdot HCl. \end{cases}$$

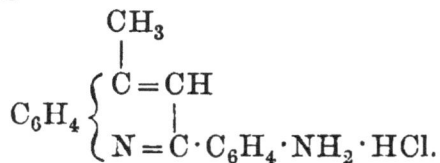

Application.—Flavaniline is a basic colouring matter, and is applied in dyeing like the other basic coal-tar

* Phenol yields a yellow, resorcin an orange, and β-naphthol a red.

colours. Wool and silk are dyed in neutral baths; cotton is previously prepared with tannin and tartar emetic. The shades obtained are bright yellow with a slight cast of green. On silk the colouring matter shows a beautiful moss-green fluorescence.

Flavaniline S is the sulphonic acid of flavaniline.

Phosphine Aniline-orange, Chrysaniline, Philadelphia-yellow (impure).

Preparation.—In the manufacture of phosphine, the resinous bye-products obtained in the manufacture of magenta by the arsenic-acid process are used. These contain, as has already been mentioned, violaniline, mauvaniline and chrysaniline, a little rosaniline, and bodies of a resinous character.

To separate these, a very tedious process has to be adopted. It consists in first boiling the mass with dilute hydrochloric acid. The violaniline and resins remain insoluble. By fractional precipation, first the mauvaniline, then the rosaniline, and finally chrysaniline, are thrown down.

Another source of chrysaniline are the arsenical mother-liquors which remain after precipitating the magenta with salt. Salt and lime are added to these, the precipitate is dissolved in nitric acid, and the sparingly soluble nitrate of chrysaniline precipitated with excess of nitric acid, and then converted into the hydrochloride.

A base of higher molecular weight, corresponding to chrysaniline, is obtained by heating arseniate of toluidine to 130°–150°, and is known as chrysotoluidine.

Properties.—Phosphine, $C_{19}H_{15}N_3 \cdot HNO_3$, is the nitrate of chrysaniline or diamidophenylacridine :—

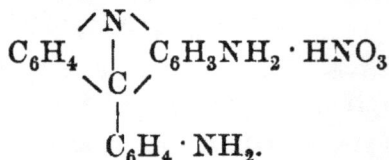

$$C_6H_4 \begin{array}{c} \diagup N \diagdown \\ | \\ \diagdown C \diagup \\ | \end{array} C_6H_3NH_2 \cdot HNO_3$$

$$C_6H_4 \cdot NH_2.$$

It comes into commerce as a yellow or orange powder, easily soluble in water and alcohol.

If concentrated hydrochloric acid is added to an aqueous solution of phosphine, a precipitate of the diacid salt is produced, which is easily soluble in water.

The reaction of phosphine solutions with nitrates is especially characteristic. If a solution of saltpetre containing 1 part per 100 is added, a red crystalline precipitate of nitrate of chrysaniline is immediately produced. In warm solutions the precipitate is only produced on standing ; under the microscope it is seen to consist of needles.

In solutions of phosphine, ammonia, and caustic alkalies precipitate chrysaniline, $C_{20}H_{17}N_3 \cdot H_2O$, as an amorphous yellow powder, soluble in alcohol and ether.

Phosphine is easily reduced by stannous chloride and hydrochloric acid, but the colour is quickly reproduced on exposure to air. This reaction is useful in the detection of phosphine.

Application.—Phosphine is not much used, but finds some application in silk-dyeing and in woollen-printing.

Silk may be dyed in a pure soap bath, and the colour brightened with acetic acid. A mixture of magenta and phosphine gives a fine scarlet on silk. In wool-dyeing an addition of acid is necessary.

On cotton, mordanted with aluminium acetate (red liquor) it gives a nankin-yellow, which resists the action of soap.

Detection on the fibre.—Acids redden the colour, but on standing it is stripped. Ammonia turns it to a greenish yellow, which is lighter than the original colour. Stannous chloride and hydrochloric acid decolourise it gradually.

IV. Artificial Indigo.

On the 1st of May, 1880, a patent was taken out by Professor Baeyer, of Munich, for the "Preparation of Derivatives and Homologues of Ortho-nitro-cinnamic Acid, and their conversion into Indigo-blue and allied Dye-stuffs." Thus indigo-blue, or indigotine, belongs to the coal-tar colours. In spite of the endeavours of the "Badische Anilin und Soda Fabrik," this synthesis has not yet become a practical success, since the artificial indigo is much more expensive than the natural product. If the cost of artificial indigo could be reduced below that of natural indigo, its introduction into commerce would cause as great a revolution in dyeing as that of alizarin in place of madder.

Of more practical importance than artificial indigo is the "propiolic acid," which, in itself colourless, may be converted into indigo-blue on the fibre.

Propiolic Acid.

(Ortho-nitrophenyl-propiolic Acid.)

Preparation.—The starting-point for the manufacture is ortho-nitro-cinnamic acid—

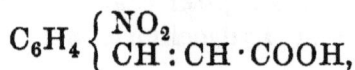

$$C_6H_4 \begin{cases} NO_2 \\ CH:CH\cdot COOH, \end{cases}$$

which is obtained from cinnamic acid.

Cinnamic acid, $C_6H_5 - CH:CH - COOH$, may be obtained by heating benzaldehyde with acetic anhydride :—

$$2C_6H_5COH + \begin{matrix} CH_3 \cdot CO \\ CH_3 \cdot CO \end{matrix} \diagdown_{\diagup} O =$$

Benzaldehyde. Acetic anhydride.

$$2C_6H_5 - CH:CH - COOH + H_2O.$$

Cinnamic acid.

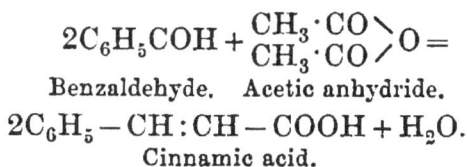

On a [large scale, benzal chloride is heated to 180°–200° with anhydrous sodium acetate :—

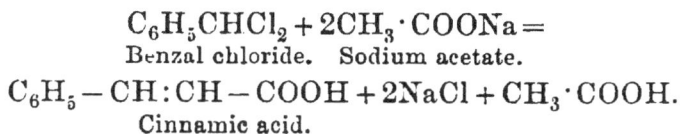

$$C_6H_5CHCl_2 + 2CH_3 \cdot COONa =$$

Benzal chloride.   Sodium acetate.

$$C_6H_5 - CH:CH - COOH + 2NaCl + CH_3 \cdot COOH.$$

Cinnamic acid.

The cinnamic acid is treated with fuming nitric acid, which converts it into two isomeric mononitro acids. The ortho acid (which alone is used) is separated from the para acid by a method based upon the different solubility of the two products.

Another method is to heat ortho-nitro benzaldehyde with acetic anhydride :—

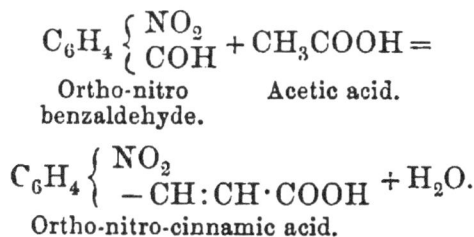

$$C_6H_4 \begin{cases} NO_2 \\ COH \end{cases} + CH_3COOH =$$

Ortho-nitro   Acetic acid.
benzaldehyde.

$$C_6H_4 \begin{cases} NO_2 \\ -CH:CH \cdot COOH \end{cases} + H_2O.$$

Ortho-nitro-cinnamic acid.

Ortho-nitro-cinnamic acid unites with bromine to form a dibromide :—

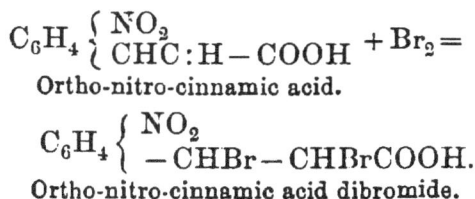

$$C_6H_4 \begin{cases} NO_2 \\ CHC:H - COOH \end{cases} + Br_2 =$$

Ortho-nitro-cinnamic acid.

$$C_6H_4 \begin{cases} NO_2 \\ -CHBr - CHBrCOOH. \end{cases}$$

Ortho-nitro-cinnamic acid dibromide.

This dibromide is boiled with alcoholic potash :—

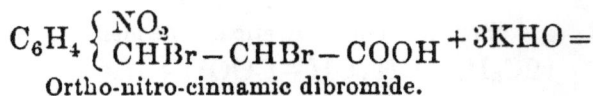

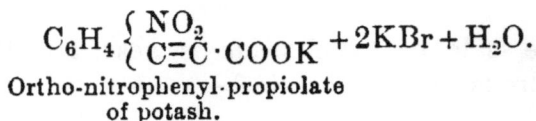

$$C_6H_4 \begin{cases} NO_2 \\ CHBr-CHBr-COOH \end{cases} + 3KHO =$$
Ortho-nitro-cinnamic dibromide.

$$C_6H_4 \begin{cases} NO_2 \\ C\equiv C \cdot COOK \end{cases} + 2KBr + H_2O.$$
Ortho-nitrophenyl-propiolate
of potash.

The free acid is separated by means of hydrochloric acid.

From ortho-nitrophenyl-propiolic acid and its ethers certain intermediate products may be obtained. These are indogenic acid and its ethers, and indogene, both of which are converted by dilute acids and alkalies, in presence of air, into indigo-blue (not indoïn), thus being suitable for calico-printing.

Properties.—The pure acid forms colourless needles or leaflets, which become brown on heating, and swell up and decompose at 155° to 160°. It is soluble in boiling water, but is decomposed by long boiling. Its alkali salts are easily soluble in water, but are precipitated by excess of alkali.

The acid comes into commerce as " propiolic acid," in a paste containing 25 per cent. of solid substance.

By boiling with reducing agents, such as potash and grape-sugar, indigo-blue is formed :—

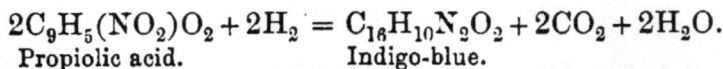

$$2C_9H_5(NO_2)O_2 + 2H_2 = C_{16}H_{10}N_2O_2 + 2CO_2 + 2H_2O.$$
Propiolic acid.   Indigo-blue.

If propiolic acid, in sulphuric-acid solution, is treated with reducing agents, like ferrous sulphate, metallic iron, zinc, tin, lead, copper, etc., or is reduced with sulphurous acid, potassium sulphocyanide, etc., indigo-blue is not produced, but a very similar colouring matter, known as indoïn, is formed. If, for instance, propiolic acid is mixed with sulphuric acid and ferrous sulphate, the solution turns blue and indoïn separates as a flocculent precipitate. Indoïn may be distinguished from indigo by the following reactions :—

It dissolves in cold sulphuric acid, but does not easily yield a sulphonic acid on heating. With sulphurous acid or bisulphites, a blue solution is produced, from which a colouring matter soluble in water may be salted out. On warming, this double compound is decomposed, yielding sulphurous acid and the original colouring matter.

Application.—Propiolic acid is used exclusively for calico-printing.

The printer's colour consists of a solvent, a reducing agent, and a thickening. The most suitable combination is borax, xanthate of soda or zinc, and starch.

The colour does not stand steaming well. It is developed like aniline-black. In this operation the unpleasant smell of mercaptan, C_2H_5SH, is produced, which adheres very tenaciously to the material, but can be removed by treatment with weak alkalies and soap-baths.

The blue obtained in this manner is not pure, but slightly grey.

The production of propiolic-acid blue is thus very simple. On the other hand, the production of blue with natural indigo is very complicated, as the following example will show :—

A tin-indigo compound is first prepared by reducing indigo with stannous chloride and soda, and precipitating with hydrochloric acid. ' This indigo-white compound is thickened, and printed, and the pieces passed through milk of lime, which dissolves the indigo-white and allows it to penetrate the fibre. They are then immersed in running water, which effects the oxidation and the fixation of the blue. Then follows an acid bath to remove the lime, and finally a soap bath. The colour is never very intense, and the operation must be carried out with great care and rapidity, in order to prevent the

U

indigo-white becoming oxidised before it has had time to be absorbed by the fibre.

Propiolic-acid blue, however, possesses the disadvantage that it cannot be used along with other colouring matters which require steaming, since it is turned grey by this operation.

It may be printed along with aniline-black, and also with iron and alumina mordants, which are afterwards dyed in alizarin.

Detection on the fibre.—It is unaltered by hydrochloric acid and caustic soda. Stannous chloride first turns it green, and then decolourises it. When warmed with an alkaline solution of hyposulphite of soda, the colouring matter is first reduced and then removed from the fibre. If this solution is exposed to the air, it becomes blue, the indigo being regenerated. Concentrated nitric acid turns it yellow. Boiling alcohol is not coloured, but chloroform becomes blue.

On heating a small piece of the material purple vapours are given off.

V. The Anthracene Colouring-Matters.

Although the number of colouring matters obtained from anthracene is very limited, they all resemble each other so closely in their chemical properties, and at the same time differ so widely from the other coal-tar colours, that they form a natural group of the latter.

In the free state they are almost insoluble in water, but are easily soluble in ammonia and caustic alkalies. With the alkaline earths and most metallic oxides they yield richly coloured, insoluble lakes. This behaviour towards bases indicates the acid nature of these colouring matters, and a more thorough examination will show that they contain free hydroxyl groups, and therefore belong to the phenols.

The anthracene colouring-matters can only be dyed with the help of mordants; they are invariably adjective. The shades produced with them are much faster to soap, chloride of lime, dilute acids, and in most cases also to light, than those obtained with the other coal-tar colours, while at the same time they are faster than most of the natural colouring matters.

Manufacture of Commercial Alizarin.

The method usually adopted for the transformation of anthracene into alizarin is effected in three distinct operations. Anthracene is first oxidised to anthraquinon, which is then treated with sulphuric acid, and thereby transformed into sulphonic acids. These are melted in the third operation with caustic soda, when alizarin is formed. The following anthracene derivatives are formed during the process :—

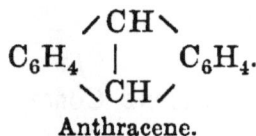

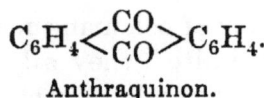

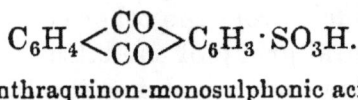

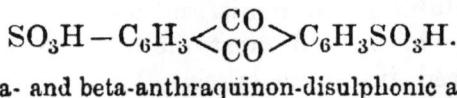

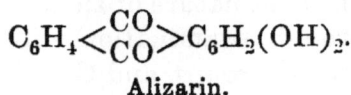

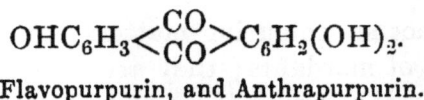

$$C_6H_4 \begin{matrix} CH \\ | \\ CH \end{matrix} C_6H_4.$$

Anthracene.

$$C_6H_4 < \begin{matrix} CO \\ CO \end{matrix} > C_6H_4.$$

Anthraquinon.

$$C_6H_4 < \begin{matrix} CO \\ CO \end{matrix} > C_6H_3 \cdot SO_3H.$$

Anthraquinon-monosulphonic acid.

$$SO_3H - C_6H_3 < \begin{matrix} CO \\ CO \end{matrix} > C_6H_3SO_3H.$$

Alpha- and beta-anthraquinon-disulphonic acids.

$$C_6H_4 < \begin{matrix} CO \\ CO \end{matrix} > C_6H_2(OH)_2.$$

Alizarin.

$$OHC_6H_3 < \begin{matrix} CO \\ CO \end{matrix} > C_6H_2(OH)_2.$$

Flavopurpurin, and Anthrapurpurin.

In order to prepare alizarin, anthrapurpurin, or flavopurpurin from this product, it is necessary to introduce two or more hydroxyl groups; and this is effected by means of a well-known reaction based upon the behaviour of the sulpho groups towards molten caustic alkali, which is also used, for instance, in the manufacture of resorcin. In melting the anthraquinon-sulphonic acids with alkali, however, the sulpho groups are not simply replaced by hydroxyl, but an oxidation invariably takes place simultaneously.

Thus when anthraquinon-monosulphonic acid is melted with caustic soda in the manufacture of alizarin, the reaction does not take place according to the equation—

$$C_{14}H_7O_2 \cdot SO_3Na + 2NaHO =$$
$$C_{14}H_7O_2ONa + Na_2SO_3 + H_2O,$$
Oxyanthraquinon,

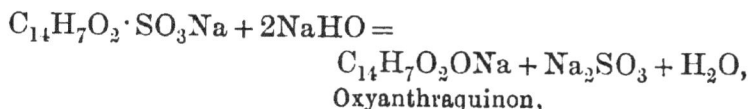

but according to the following equation :—

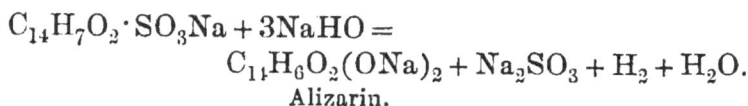

$$C_{14}H_7O_2 \cdot SO_3Na + 3NaHO =$$
$$C_{14}H_6O_2(ONa)_2 + Na_2SO_3 + H_2 + H_2O.$$
Alizarin.

The hydrogen formed according to this equation does not make its appearance in the gaseous condition, but acts as a reducing agent on the anthraquinon-monosulphonic acid and on the alizarin formed. This would diminish the yield considerably; but it is checked by adding to the melt some oxidising substance, such as chlorate of potash.

Melting with caustic soda.—In this operation anthraquinon-monosulphonic acid yields alizarin; the alpha-disulphonic acid, flavopurpurin; and the beta acid, anthrapurpurin.

The operation is carried out under a high pressure in strong boilers provided with stirring gear. From 3 to 4 parts of solid caustic soda are first heated with a small quantity of water, until the whole becomes liquid, after which the chlorate of potash is added along with 1 part of anthraquinon-monosulphonate of soda. The boiler is then closed, and the contents heated for twenty-four hours to 180°–200°. The melt is then allowed to cool, and is dissolved in a large quantity of water and neutralised with hydrochloric acid. According to a more recent patent, this neutralisation is effected with sulphurous acid, in order to effect a regeneration of the caustic soda. The alizarin is precipitated in a very finely divided flocculent state. The whole is now pumped through filter-presses, well washed, and the cakes of alizarin obtained in this manner ground with water to a homogeneous paste containing a known percentage. The latter contains alizarin, flavopurpurin, and anthrapurpurin in varying proportions.

Alizarin, $C_{14}H_6O_2(OH)_2$.

For the preparation of pure alizarin, *pure* anthraqui-non-monosulphonic acid is melted with caustic soda. It can also be obtained from the blue shade of commercial alizarin in the following manner:—

The paste is dissolved in dilute caustic soda, and the solution is filtered from the anthraquinon and other in-soluble impurities. A solution of barium chloride is then added, and the whole is heated to the boil, when alizarate of barium separates out in a crystalline state. The precipitate is collected on a filter, washed, and decom-posed with an acid. The washing is then continued, in order to remove the barium and free acid. The product obtained in this manner is almost chemically pure, but it can be purified still further by sublimation or crystal-lisation from glacial acetic acid.

Alizarin has also been obtained synthetically by heating phthalic anhydride with pyrocatechin:—

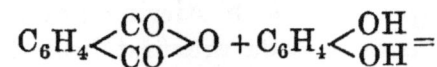

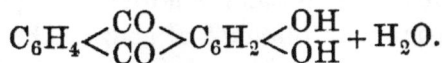

$$C_6H_4 {<}^{CO}_{CO}{>}O + C_6H_4{<}^{OH}_{OH} =$$

Phthalic anhydride.   Pyrocatechin.

$$C_6H_4{<}^{CO}_{CO}{>}C_6H_2{<}^{OH}_{OH} + H_2O.$$

Alizarin.

The position of the hydroxyl groups is shown in the following constitutional formula:—

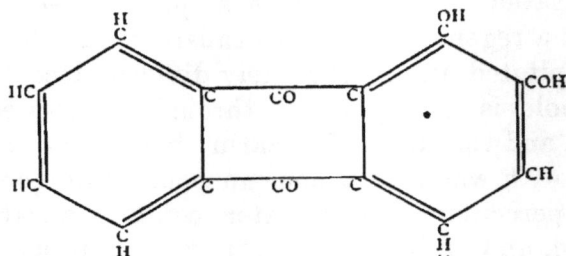

Sublimated alizarin forms splendid orange-red crystals, which melt at about 280°. It is almost insoluble in cold water and but sparingly soluble in boiling water, one litre of which takes up 0·31 grm. of alizarin. In cold alcohol it does not dissolve easily, but boiling alcohol and glycerin dissolve it. Sulphuric acid dissolves alizarin with a reddish-brown colour, but on diluting with water the alizarin is reprecipitated in the unchanged state. In solutions of alum or aluminium sulphate alizarin is almost insoluble.

Alizarin crystallised from moist ether contains three molecules of water of crystallisation. When precipitated by acids from alkaline solutions, it also contains water, which is given off at 100°.

The behaviour of alizarin towards alkalies is that of a weak acid. It dissolves in caustic alkalies and in ammonia with a blue-violet colour, but is precipitated from these solutions by acetic acid. If, however, alizarin is boiled with sodium acetate, it dissolves and is separated out unchanged on cooling. But if the boiling is continued for some time, acetic acid is given off and the alizarin remains permanently dissolved.

The compounds of alizarin with calcium and barium are thrown down as violet precipitates when salts of lime or baryta are added to an alkaline solution of the colouring matter. The peculiar behaviour of the calcium compound towards free carbonic or acetic acids is described on p. 304. (Rosenstiehl's method of dyeing.)

With all other bases alizarin yields sparingly soluble or insoluble permanent lakes. The alumina and tin lakes are red, the others mostly dark-coloured. The behaviour of alizarin towards these bases is that of a strong acid; so strong, indeed, that it is capable of decomposing chlorides and nitrates in dilute solutions, thus replacing hydrochloric and nitric acids.

Anthrapurpurin.

Pure anthrapurpurin (also called isopurpurin) is best obtained by melting pure beta-anthraquinon-disulphonic acid with caustic soda and chlorate of potash. Nothing more is known of the position of the hydroxyl groups than that they are contained in both benzene rings :—

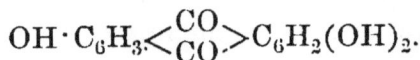

$$OH \cdot C_6H_3 {<}^{CO}_{CO}{>} C_6H_2(OH)_2.$$

Anthrapurpurin sublimates with partial decomposition, and melts at 360°. It is somewhat soluble in boiling water; easily soluble in boiling alcohol and glacial acetic acid.

Its solutions in caustic alkalies are redder than those of alizarin. It is somewhat soluble in a solution of barium hydrate. Sulphuric acid dissolves it with a dull violet colour, but the addition of a trace of sodium nitrite to the solution turns it a splendid red-violet. A boiling solution of alum dissolves a trace of anthrapurpurin with an orange colour, which on cooling separates out again. (Distinction from purpurin.)

Flavopurpurin.

With respect to formula and preparation, the same applies as for anthrapurpurin, except that in the preparation, instead of the beta acid, alpha-anthraquinon-disulphonic acid is used. Flavopurpurin forms yellow needles, which melt above 330° and can be sublimated It is almost insoluble in water, but dissolves easily in cold alcohol.

Its solutions in alkalies are redder still than those of anthrapurpurin; the solution in ammonia is yellowish-red, while that of anthrapurpurin is violet. A boiling solution of alum does not dissolve it.

Purpurin.

Purpurin, $C_{14} H_5 O_2 (OH)_3$, is isomeric with anthrapurpurin and flavopurpurin, and was the first trioxyanthraquinon known. Although it is not a constituent of artificial alizarin, it will be described here, since after alizarin it is the most important constituent of madder, and the results obtained in dyeing with artificial alizarin and with madder are often compared with each other.

Purpurin differs from its two isomers in having all the three hydroxyl groups in one benzene ring. Their relative position is shown in the formula :—

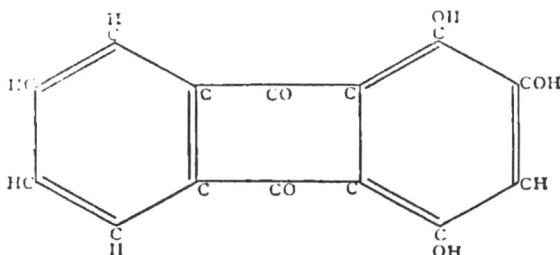

Lalande has devised a method for the preparation of purpurin from alizarin. Alizarin is dissolved in concentrated sulphuric acid, and is oxidised by the introduction of dry arsenic acid, and heating to 150°–160°. When the reaction is over, the colouring-matter is precipitated by dilution with water, and washed with a concentrated solution of alum, which dissolves the purpurin, while unchanged alizarin remains behind. The liquid is filtered and the purpurin precipitated by the addition of hydrochloric acid. Pure purpurin is obtained by crystallising the precipitate obtained in this manner from alcohol.

It forms red needles, which begin to sublimate at 150°, and melt at 253°. In boiling water it dissolves with a dark-red colour. When crystallised from aqueous alcohol, it is obtained in the form of orange needles, of the formula $C_{14}H_5O_2(OH)_3 + H_2O$.

Purpurin dissolves in caustic alkalies with a purple colour; but if the solution is exposed to the light for some time, the colouring-matter is destroyed and the liquid is decolourised.

The most characteristic property of purpurin is, that it dissolves in a boiling solution of alum with a yellowish-red colour and green fluorescence. The purpurin-alumina-lake also shows this property.

If a solution of alizarin in alkalies or alum is treated with an excess of acid, *purpurin hydrate* is thrown down, which dissolves in tepid alcohol much more easily than purpurin. When heated, it parts with its water of hydration and passes over into ordinary purpurin. The artificial purpurin-paste is supposed to consist chiefly of this hydrate.

Commercial Alizarin.

Alizarin is always sold in commerce in the form of a paste, which contains the hydrates of the colouring matters in an exceedingly fine state of division. The amount of dry colouring-matter is usually 20 per cent., but pastes are sometimes sold which contain as much as 60 per cent. of dry substance. These concentrated pastes are not so advantageous as the thinner ones, as they do not divide as evenly in the dye-baths and in printer's colours. According to an agreement lately made among the German alizarin manufacturers, no other than 20 per-cent. alizarin is brought into the market.

If alizarin-paste is dried and afterwards ground to a paste again with water, it is found to have lost its slight solubility altogether, and has become totally unfit for dyeing. The reason of this change is, that it is very difficult to obtain the original fine state of division by grinding; and besides that, the colouring matters give up their water by hydration on drying.

The two most important commercial varieties are alizarin-blue shade, or alizarin V, and alizarin-yellow shade, or alizarin G.

The blue shade of alizarin is obtained by melting crude anthraquinon-monosulphonic acid with caustic soda, and consists chiefly of alizarin. In dyeing it yields with an alumina mordant a bluish but not very brilliant shade of red; with a small percentage of mordant, however, beautiful shades of pink can be obtained. It is used besides along with an iron mordant for dyeing and printing fast violets.

Alizarin-yellow shade contains a large percentage of anthrapurpurin and flavopurpurin, and little alizarin. Anthrapurpurin yields with alumina mordants an almost neutral red; flavopurpurin, however, yields a fiery red which contains a considerable proportion of yellow. The larger the proportion of flavopurpurin the yellower the shades. The violets obtained with flavopurpurin, and anthrapurpurin, and iron mordants are of no use.

The chief impurities which alizarin-blue shade contains are anthraquinon and oxyanthraquinon.

Oxyanthraquinon, $C_6 H_4 < {CO \atop CO} > C_6 H_3 OH$, is the mono-hydroxyl derivative corresponding to anthraquinon-mono-sulphonic acid; and it would be the only product of the alizarin melt, were it not that a simultaneous oxidation takes place and alizarin is formed.

In a similar manner alizarin-yellow shade contains two dioxyanthraquinones, $C_{14} H_6 O_2 (OH)_2$, which are isomeric with alizarin.

Anthraflavic acid is produced from the alpha-disulphonic acid of anthraquinon; flavopurpurin can be regarded as an oxyanthraflavic acid. Similar relations exist between beta-anthraquinon-disulphonic acid, isoanthraflavic acid, and anthrapurpurin.

The oxyanthraquinones which occur as impurities in alizarin may be detected by boiling with lime and filtering. If the filtrate is coloured brown, the sample contains oxyanthraquinon, anthraflavic acid, or isoanthraflavic acid, all of which can be precipitated by the addition of an acid. None of these impurities are of any value in dyeing, and therefore only diminish the value of the alizarin.

The different colouring matters contained in commercial alizarin can be recognised in the following manner :—

A small sample of the alizarin in question is dissolved in sodium carbonate, and the solution filtered from the undissolved anthraquinon and oxyanthraquinon, which latter can be separated from each other by means of caustic alkali or quicklime. The filtrate is then acidified with hydrochloric acid, and the precipitate boiled with milk of lime in order to remove anthraflavic and isoflavic acids. The undissolved lime-lakes are stirred to a paste with water, decomposed with hydrochloric acid, and the residue collected on a filter, washed, and dried.

The residue obtained in this manner is a mixture of alizarin, flavopurpurin, and anthrapurpurin; but the quantitative separation of these three products is very difficult to carry out. According to the following method of Schunck and Römer, they can easily be detected when present simultaneously :—

A small sample is dried at 100°, and then placed between two glass plates separated from each other by a leaden ring some millimetres in thickness. The whole is heated in an air-bath to 140°–150°, at which temperature the alizarin sublimates and is carried away: if the temperature is now raised to 170°, a mixed sublimate of flavopurpurin and anthrapurpurin is obtained. These two colouring-matters are most readily detected under the microscope ; flavopurpurin forms reddish-yellow needles, while anthrapurpurin forms thick crystals. The·quantitative separation of the two might be effected by boiling with benzene, in which flavopurpurin dissolves, anthrapurpurin not.

The *valuation* of the alizarins is usually effected by estimating the percentage of dry substance and the ash, and by carrying out comparative dye-trials.

In estimating the percentage of dry substance, it should be borne in mind that the temperature should not be allowed to rise much above 100°, since alizarin begins to sublimate at 110.° The residue should appear yellow, and not dark brown. Alizarin pastes sometimes contain glycerin, Turkey-red oil, etc., which have been added in order to thicken the paste. These can be separated from the colouring-matters by diluting with water and filtering. The filtrate may contain besides, small quantities of salts which have not been properly removed in the manufacture. It should neither have a brown nor a reddish tinge, but should be perfectly colourless.

The ash should not weigh more than 1 per cent. of the dry alizarin, and should be free from iron.

Application of Alizarin.

The alizarin colouring-matters are adjective in their behaviour towards all textile fibres. The shades produced with the help of different mordants are characterised by their fulness, brilliancy, and fastness. They are used principally for dyeing and printing cotton, and to some extent in wool-dyeing. On silk they yield fine shades of red, brown, black, etc.; but they are not used much in silk-dyeing, on account of the great advantages which the substantive coal-tar colours possess over alizarin. Their fixation with the help of mordants not only necessitates a much longer and more difficult process of dyeing, but has besides the disadvantage of causing the silk to lose some of its gloss and pliability, on account of the alumina deposited on the fibre.

COTTON.—The mordants used in cotton-dyeing and calico-printing are salts of alumina, iron, chromium, and tin, besides lime, tannic acid, and oil mordants.

Alizarin-red and pink.—Alizarin-red is the alumina lime-lake of the yellow shade, alizarin-pink the alumina lime-lake of the blue shade of alizarin. These lakes can be obtained as such by dissolving the colouring-matter in an alkali and adding a solution of alum, in the form of red flocculent precipitates. The combination of pure alizarin with alumina is the most stable; and the affinity which these two bodies have for each other is so great that if alizarin is boiled with a dilute solution of aluminium sulphate, the latter is decomposed, aluminium alizarate being formed. An additional proof of the affinity of these two substances is rendered by the fact that alizarin is precipitated from its alkaline solutions when agitated with aluminium hydrate.

When dyed on cotton, alizarin is not present in the form of pure alumina lake, but in a much more complicated combination, which contains besides alumina two other inorganic bases—viz., lime and stannic oxide. The compound contains besides, tannic acid, fatty acids, and in some cases phosphoric and arsenic acids, which, along with the alizarin, form the acid constituents. Alizarin-red thus offers a very striking example of the rule that the stability of a colour is proportionate to the number of the different constituents of the lake.

Alizarin can be fixed on the fibre in two ways—viz., either by *dyeing* or by *steaming*. In dyeing, the material is first prepared with the mordant (alumina, iron, etc.), and then dyed in a hot bath of alizarin; while in steaming, the alizarin and mordant are printed simultaneously on the fabric and the colour developed by means of steam.

In Turkey-red dyeing the only mordant base which is applied is a salt of alumina, the lime and stannic oxide contained in the finished lake being introduced during the different operations of dyeing and clearing.

Yarn is mordanted with solutions of alum or aluminium sulphate neutralised with soda or chalk. The same mordant can be used for dyeing calico or other cotton fabrics. For the production of coloured designs on a white ground, it is necessary to print the mordant on those places which must subsequently appear coloured. In printing, a special mordant for red is commonly used, which is obtained by adding to a solution of alum a quantity of acetate of lead insufficient to effect a total double decomposition, along with some soda. The solution obtained in this manner contains basic acetate and basic sulphate of alumina. Acetate of alumina alone would not give good results, as it renders the material impervious, and difficult to wet out. The mordant is thickened with starch for printing.

The following operations effect the precipitation of the alumina imbibed by the fibre in a more or less granular but not gelatinous condition. The material is first hung up in a large chamber heated to 32°–40° C., in which the degree of moisture is carefully regulated. It remains here until a considerable proportion of the acetic acid has been given off, which renders the alumina-salts more basic and almost insoluble.

In the next operation, the " dunging," the material is passed through a hot bath containing cow's (or pig's) dung and chalk. The chalk removes acetic and sulphuric acids from the printed material, while the action of the cow's dung is chiefly a mechanical one ; it prevents any detached mordant from settling on the white parts of the fabric.

The pieces are now thoroughly washed and passed into the dye-bath, where they are dyed with alizarin, beginning at a low temperature, and raising gradually to the boil.

The analysis of materials dyed with madder, therefore, with a mixture of natural alizarin and purpurin, appears to show that the red contains alumina and lime in the proportion of two molecules of Al_2O_3 to three of CaO. According to the more recent investigations of Liechti and Suida, the most probable formula for the normal alizarin-red lake (before soaping) is:—

$$Al_2O_3 \cdot CaO(C_{14}H_6O_3)_3 \cdot H_2O.$$

The subsequent operation of soaping, however, disturbs the ratio of the lime to the alumina, some of the latter being removed. The material contains, previous to dyeing, a small quantity of lime, which has become assimilated during the bleaching process from the chloride of lime used. But the greater part of the lime is taken from the calcareous water used in preparing the dye-bath.

According to Rosenstiehl, it is necessary for this reason that the water should contain a certain amount of lime. Although the alizarin lime-lake is insoluble in pure water, it dissolves easily in carbonic acid; and as the natural waters usually contain bicarbonate of lime as well as free carbonic acid, a considerable proportion of the alizarin is dissolved as lime-lake, which is taken up as such by the alumina on the fibre. But as the dyeing is always carried out at a high temperature, carbonic acid is gradually evolved, and the lime-lake which is thus caused to fall accumulates at the bottom of the vessel, and causes a loss of alizarin. In order to counteract this evil, Rosenstiehl recommends an addition of calcium acetate to the dye-bath, or in case the water contains much lime, simply an addition of acetic acid. At first the bath is kept slightly acid. The alizarin dissolves in the calcium acetate, which it gradually decomposes as the temperature rises. Acetic acid is thus set free, while the alizarin lime-lake formed is taken up along with free alizarin by the alumina.

After dyeing, the red has a dull appearance, and must pass through the operation of *clearing* in order to obtain its full brilliancy.

This is effected by boiling with soap and stannous chloride. The action of the latter is rather complicated. In the first place, it removes part of the soda from the soap, and yields sodium chloride and stannous hydrate. The soap is thus rendered more neutral and more fitted to give up fatty acid to the colour-lake, which is proved by the considerable quantity of fatty acid taken up during this operation. The stannous hydrate reduces and thereby decolourises the brown impurities fixed on the fibre along with the alizarin. It is thereby gradually oxidised to stannic hydrate, which combines with part of the alizarin to form an orange lake, and thus brightens the colour.

According to another method, stannic oxide is introduced into the colour-lake by passing the pieces through a moderately warm bath containing nitromuriate of tin. The latter is a stannic salt, and is obtained by adding stannous chloride to an equal weight of nitric acid.

The fastness of alizarin-red is due to a great extent to the presence of *fatty acids* or of *oxidised fatty acids*. The amount of fatty acid introduced during the operation of soaping is insufficient, and special operations are therefore necessary in order to charge the fibre with this substance. Turkey-red, which is the fastest and most permanent colour obtained with alizarin, contains the maximum amount of fatty acids.

The oil which yielded the fatty acids in the old method of Turkey-red dyeing was known as *huile tournante*, or *rancid Gallipoli*. It is a species of olive oil which contains free oléic acid, and possesses the property of forming an emulsion with alkaline carbonates. The yarn or fabric

x

was passed through, or padded, with an emulsion of this kind and then exposed to the air, whereby part of the oil was oxidised to a compound insoluble in alkaline carbonates. The operation required, however, a great deal of time and trouble.

At present rancid Gallipoli is not used at all, or only in a few special cases. It has been replaced by the so-called Turkey-red oil (see p. 51), with which the same results can be obtained more simply and more rapidly. The operation of *oiling* can either take place before mordanting with alumina, or after dyeing, in which latter case the action is assisted by steaming. In some cases Turkey-red oil is added to the dye-bath.

Sumach, or some other form of tannin, is usually added to the dye-bath. It combines with part of the alumina, and thus yields fuller shades. Glue is frequently added besides, in order to render the tannic acid insoluble, and thus to prevent it from becoming attached to those parts which must remain white. On heating, the alumina on the fibre decomposes the finely divided compound of tannic acid and glue, and becomes saturated with tannic acid.

In printing it is often desirable to use a strongly alkaline mordant in place of the ordinary mordant for red. Aluminate of soda is generally used for this purpose. The fixation of the mordant is effected by means of a weak solution of ammonium chloride, according to the equation :—

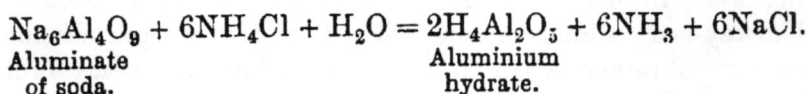

$$Na_6Al_4O_9 + 6NH_4Cl + H_2O = 2H_4Al_2O_5 + 6NH_3 + 6NaCl.$$

Aluminate   Aluminium
of soda.   hydrate.

From this it follows that the alumina is not thrown down in the form of ordinary gelatinous aluminium hydrate, $Al_2(OH)_6$, but in a more granular condition, containing less water of hydration. The printer's colours for steam reds and pinks invariably contain, besides thickening, alizarin and a salt of alumina, acetate of lime. The salt of alumina must be one which will be decomposed by the heat of the steaming chamber, and usually consists of acetate or sulphocyanide of alumina.

The acetate of alumina (prepared from acetate of lead and alum, or by dissolving aluminium hydrate in acetic acid) is decomposed on heating, and the acetic acid liberated helps to dissolve the alizarin, and thus enables it to penetrate the fibres and combine with the alumina. The acetate of lime also helps to dissolve the alizarin, and yields besides the lime necessary for the formation of the alizarin-lake. When acetate of lime is used, the acetate of alumina is usually replaced by nitrate of alumina; the colours are thus rendered brighter.

The slightest trace of ferric oxide suffices to change the fine red or pink shades of the alumina-lake to a dull red or even a brown. The printer must therefore take special precautions to keep his colour free from iron. But as the latter is in continuous contact with the steel "doctors" of the printing-machine during the operation of printing, traces of iron are always taken up by the colour. Now, the presence of small quantities of iron is no disadvantage as long as it is prevented from entering into the alizarin-lake in the ferric state, and this can be prevented in several ways.

Thus an addition of stannous chloride or stannous hydrate reduces all the iron to the ferrous state. Arsenious acid is more effective; it combines with the ferric oxide to form an insoluble salt, which is not decomposed by alizarin. An addition of sulphocyanide of potassium to the colour is very serviceable. The iron is oxidised to the ferric state by the nitrate of alumina, and then combines with the sulphocyanide of potassium to form ferric sulphocyanide. In some cases sulphocyanide of potassium is dispensed with, and the whole of the alumina contained in the colour replaced by sulphocyanide of alumina.

The colours containing sulphocyanide of alumina are at the same time excellent resists under other colours which contain strong acids or oxidising agents. If, for instance, alizarin-red is printed and steamed, it can be printed over with aniline-black (of course without being previously washed). The oxides of chlorine which are evolved during the development of the black act on the hydro-sulphocyanic acid and on the surplus of sulphocyanide of alumina, and form persulphocyanogen, while the red remains intact.

After steaming, the material must be passed through a slightly alkaline bath, containing chalk, soluble glass, etc., in order to remove the thickening, the acetic acid produced in steaming, as well as any colour-lake not attached to the fibre, etc. The material can then be soaped, oiled, cleared, etc.

Alkaline solutions of alizarin produce a violet precipitate with ferrous salts, while with ferric salts they yield an almost black precipitate.

The production of *alizarin-violet* on cotton is quite similar to the process of Turkey-red dyeing. The blue shade of alizarin is used for this purpose; the yellow shade cannot be used, since anthrapurpurin yields greyish-violet shades, while flavopurpurin yields a reddish violet.

Ferrous acetate is usually employed as a mordant. It is decomposed in the drying chamber into acetic acid and ferrous hydrate. The material is then passed through cow's dung and chalk, after which it is washed and dyed with alizarin. Arsenite of soda is frequently added to the dye-bath, in order to produce faster and brighter shades. It reduces part of the ferric oxide fixed on the fibre to ferrous oxide, and thus gives rise to a compound lake containing ferric and ferrous oxides. The ferrous oxide may be supposed to play the same part in the violet as lime in the red. Arsenious acid also enters into combination in the colour.

The steam colours for fast violet contain ferrous acetate, acetate of lime, alizarin, and a thickening. Arsenite of soda is frequently added besides.

Brown.—Mixtures of iron and alumina mordants produce fine shades of brown. Alizarin-red can also be changed to a brown by an admixture of Prussian-blue. In order to produce a brown in this manner, ferrocyanide of potash is mixed with the colour, which is decomposed in steaming by the action of the free acid, and yields Prussian-blue.

Puce.—A very fine steam puce can be produced with nitro-acetate of chromium as a mordant.

WOOL.—Alizarin, anthrapurpurin, and flavopurpurin can be used with advantage in wool-dyeing when it is necessary to produce shades fast to light and milling. They may all be applied in a similar manner and yield similar shades. The colours produced may be varied considerably according to the mordant used.

Wool mordanted with potassium bichromate and dyed in alizarin yields different shades of maroon, which are characterised by their fulness of colour and their " bloom." If in place of alizarin alone mixtures of alizarin and other artificial or natural colouring-matters are used, a great variety of shades can be obtained. Wool may also be dyed by boiling it with alizarin and bichromate in one bath.

Wool mordanted with alum and tartar yields on dyeing with alizarin, or better with anthrapurpurin, a very fine red or scarlet. The addition of chalk or calcium acetate to the dye-bath is essential in order to obtain bright colours. It is also necessary, in order to obtain even colours, to dye very gradually ; first cold, then raising the temperature slowly to the boiling-point. No good results have hitherto been obtained with alizarin and alum in one bath.

Stannous chloride and oxalic acid used as a mordant

yields shades which range from orange to a yellow shade of scarlet. Good results are obtained with stannous chloride, oxalic acid and alizarin in one bath. The colours obtained with tin as a mordant, although fast to light, are rendered dull by milling with soap.

By mordanting the wool with copperas and tartar dull violets are obtained. If much colouring matter is used, the violet is so dark as to be almost a black. Alizarin can also be dyed with copperas and oxalic acid in one bath.

Nickel ammonium sulphate and uranium acetate have also been proposed as mordants for alizarin on wool.

SILK.—Alizarin and allied colouring-matters may also be applied to silk mordanted with various metallic salts. After dyeing, the silk is usually brightened by boiling it in a soap-bath. The colours obtained are full and fast, but are only applied in exceptional cases, where fastness is an absolute necessity.

Regeneration of alizarin from spent baths.—It has been mentioned above that the fabrics printed with alizarin are passed through a chalk-bath after steaming. In this operation the surplus alizarin is removed from the fabric, and settles at the bottom of the bath in the form of a lake, or lakes, along with the chalk and impurities. In order to regenerate the alizarin, the mud is collected, dissolved in dilute hydrochloric acid, the residue washed with water, dissolved in dilute soda, filtered and precipitated again with hydrochloric acid.

If the colour contains iron, the alizarin may be regenerated by treating with warm, dilute sulphuric acid, which dissolves all the impurities, but leaves the alizarin behind.

Reactions of Alizarin when Dyed on Textile Fabrics.

Fabrics dyed or printed with alizarin do not part with any of their colour when boiled with solutions of *caustic alkalies,* moderately concentrated. Dilute *acids* have like-wise no action. Concentrated acids, however, decompose the colour-lakes and simultaneously remove the metallic base either wholly or partially. The different alizarin colours do not possess the same power of resisting the action of acids ; thus the violet is decomposed more easily than ordinary red, while Turkey-red offers the greatest resistance.

In concentrated sulphuric acid cotton fabrics dissolve at the ordinary temperature along with the alizarin colours fixed to the fibre. If the solution thus obtained is diluted with water, the alizarin is thrown down as a flocculent precipitate, which may be collected in a filter, washed, dried, and sublimated, or at least recognised as alizarin by the violet colour it imparts to solutions of the alkalies.

Some *organic,* non-volatile *acids* are known to possess the property of preventing the precipitation of the sesqui-oxides (ferric oxide, alumina, etc.) from their solutions by means of alkalies or ammonia. These acids have a comparatively strong action on the alizarin colours. If, for instance, Turkey-red cloth is printed with oxalic acid and steamed, the alumina-lake is partially decomposed, and a pink design on a red ground is the result.

Strong acid *oxidising agents,* like nitric acid or ferric chloride, destroy alizarin.

A dilute solution of *chloride of lime* has scarcely any action on Turkey-red, but it gradually destroys the ordinary alizarin-red. Turkey-red will, however, not withstand the simultaneous action of chloride of lime and an acid.

This property has found some application in the production of white designs on a Turkey-red ground.

One method, for instance, of producing designs of this kind is to print the red fabric with tartaric acid, and then pass it through a solution of chloride of lime. Wherever tartaric acid has been printed the colouring-matter is destroyed by the action of the chlorine liberated, while at the same time the alumina is dissolved by the excess of tartaric acid.

Potassium bichromate does not attack alizarin, but free chromic acid destroys it.   White designs or discharges can therefore also be produced with this compound by passing the fabric through a solution of the bichromate, drying, and subsequently printing with tartaric (or oxalic) acid, which liberates chromic acid and simultaneously dissolves the mordant.

Red prussiate of potash in alkaline solution, otherwise one of the most powerful oxidising agents, is without action on the colours produced with artificial alizarin. Dilute solutions of *permanganate of potash* are likewise without action.

Nitrous fumes convert alizarin-red into alizarin-orange.

The alizarin colours are very fast to *light* and *air*.   The red is, however, temporally affected by *heat*.   When the fabrics are dried in the usual manner, over cylinders heated by steam, it parts with some of its brilliancy and acquires a slight cast of brown.   The original colour is, however, almost completely restored by exposing the goods to the air.   It is probable that the alizarin-red lake contains some water of hydration, which is partly given off in drying, but is taken up again in a moist atmosphere.

Detection on the fibre.—In order to detect the presence of alizarin in red, brown, violet, or black cotton fabrics, the above reaction with concentrated sulphuric acid will suffice.   But in case the sample of material is too small, or the colours too light (containing only a small percentage of alizarin), the following characteristic reactions are resorted to :—

Ammonia and soda have no action on the colour, neither has a dilute solution of chloride of lime.   The material is decolourised by boiling it with a mixture of two parts alcohol and one part concentrated hydrochloric acid. This latter reaction is of value in distinguishing between alizarin-black and logwood-black, which latter is destroyed even by dilute acids, colouring the solution red. It also serves to distinguish between alizarin-black and aniline-black, which latter is not affected, or at least only turned to a greenish shade.

Alizarin-red is turned violet when boiled with barium hydrate.

Materials dyed with alizarin can be distinguished from those dyed with madder or with preparations of madder by boiling them with a solution of aluminium sulphate. The latter yield a fluorescent solution (see Purpurin, p. 295).

The spectroscope may render good services in the detection of alizarin and alid colouring-matters, as they all give in ammoniacal solution (aqueous or alcoholic) characteristic bands of absorption.

Alizarin is used at present pretty largely in wool-dyeing. But as it is seldom used as a self-colour, and is nearly always mixed with one or more natural or artificial colouring-matters, its detection is often very difficult and tedious.

Application of Purpurin.

Almost all colours which are at present obtained with artificial alizarin were formerly produced with madder, or preparations of madder. The latter contains very little anthra or flavopurpurin, but in place of these we find purpurin.

It might therefore be inferred that a mixture of the blue shade of alizarin and Lalande's artificial purpurin would yield the best substitute for madder. Nevertheless, purpurin is not used in the large scale. When used in the form of the crude product, it gives an inferior steam

red verging into a brown.   But in the purified state it is
said to yield a most brilliant scarlet.

The colours produced with artificial alizarin are faster
than those produced with purpurin.   The latter are
destroyed by alkaline solutions of red prussiate and by a
1 per cent. solution of permanganate.

In order to estimate the amount of pure purpurin in a
sample of the commercial product, a weighed quantity is
boiled with a solution of aluminium sulphate, filtered from
the undissolved alizarin, etc., and the purpurin precipitated in the filtrate by the addition of an acid.   The
precipitate is collected in a filter, washed, dried and
weighed.

For the detection of small quantities of alizarin in a
sample of purpurin an alkaline solution is prepared and
exposed to the light until all the purpurin is destroyed.
The unchanged sodium alizarate is then decomposed with
dilute sulphuric acid, the liquid extracted with ether,
evaporated to dryness, and the residue tested in the
usual way for alizarin.

Alizarin-carmine.   Alizarin S or W S.

The preparation of an alizarin sulphonic acid is effected
by acting on 1 part of alizarin with 3 parts of concentrated
sulphuric acid containing 20 per cent. of anhydride, at
100°–150° C.   The heating is continued until a sample
dissolves completely in water,   The product is then
dissolved in water and the excess of sulphuric acid
precipitated by barium hydrate or milk of lime.   The
filtrate is neutralised and evaporated down.

The product of the reaction is alizarin monosulphonic
acid, $C_{14}H_5O_2(OH)_2SO_3H$.   The free acid is soluble in
water and yields three series of salts.   Those of the
general formula, $C_{14}H_5O_2(OH)_2SO_3M$, are yellow or
orange, and are soluble in water.   The sodium-salt,
$C_{14}H_5O_2(OH)_2SO_3Na$, is the commercial product; it is

decomposed by sulphuric acid, but not by hydrochloric acid.

The alkali-salts corresponding to the general formula, $C_{14}H_5O_2(OH)(OM)SO_3M$, are reddish violet; those of the alkaline earths, reddish yellow.

The salts of the general formula, $C_{14}H_5O_2(OM)_2SO_3M$, are the most easily soluble, and are coloured intensely violet.

In spite of the advantages which alizarin-carmine presented as a strongly acid colouring-matter, soluble in water, it has nevertheless not until recently met with much success in wool-dyeing. The free acid does not dye wool directly, and it is necessary in order to obtain a red to mordant with alumina. A good scarlet can be obtained in one bath with alizarin S, alum, and oxalic acid.

Tin mordants yield orange shades.

The shades obtained on wool with chromium and iron mordants resemble those produced with ordinary alizarin.

Anthracene-brown.   Anthragallol.

Anthragallol, $C_{14}H_5O_2(OH)_3$, is formed by heating gallic acid either with phthalic anhydride and zinc chloride or with benzoic acid and strong sulphuric acid. It is isomeric with anthrapurpurin, flavopurpurin and purpurin, and its constitution is represented by the formula—

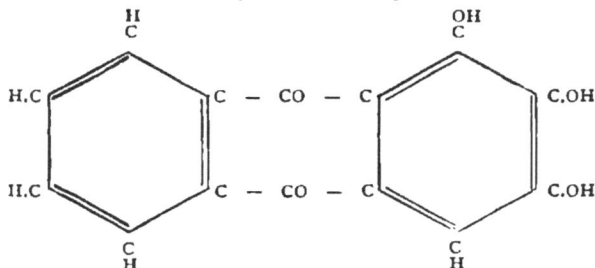

The commercial product forms a dark-brown paste, which dissolves in caustic alkalies with a greenish-blue colour.

Anthragallol yields with the usual metallic mordants various shades of brown. On wool mordanted with bichromate of potash a fine brown is obtained, which is characterised by its great fastness both to light and to milling.

Alizarin-orange.

Hitherto, two mono-nitro derivatives of alizarin have been produced, which are distinguished as α and β-nitro-alizarin. The β compound is the chief constituent of commercial alizarin-orange.

For the preparation of this colouring-matter, alizarin (blue shade) is treated in solution or in a finely divided state with nitrous fumes. The raw product is purified by dissolving it in sodium carbonate, filtering and precipitating again with an acid.

If β-nitro-alizarin is reduced, an amido-alizarin is formed, which, when treated with nitrous acid, can be transformed into purpurin $C_{14}H_5O_2(OH)_3$. The constitutional formula of β-nitro-alizarin must therefore be—

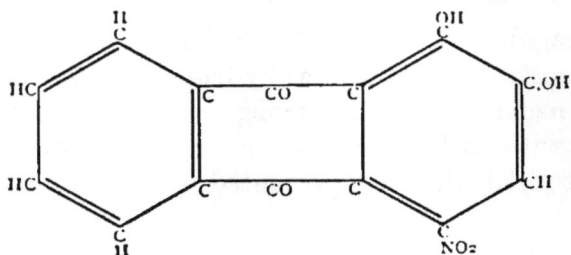

β Nitro-alizarin.

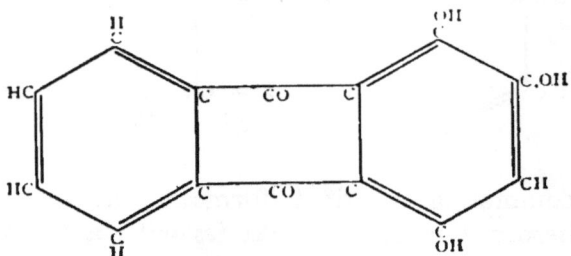

Purpurin.

Pure nitro-alizarin forms yellow needles or leaflets, which melt at 244°. At a higher temperature it sublimates with partial decomposition. It possesses stronger acid properties than alizarin.

Nitro-alizarin is almost insoluble in water, but will dissolve in glacial acetic acid. In concentrated sulphuric acid it dissolves with a golden-yellow colour.

The alkali-salts dissolve in water with a red colour, but they are insoluble in strong alkaline lyes. Nitro-alizarin can therefore only be dissolved in dilute alkalies. The other metallic salts are insoluble in water.

Alizarin-orange is brought into commerce in the form of a paste containing 10–20 (usually 15) per cent. of dry substance, or as an orange-coloured powder. The latter consists of the acid sodium salt $C_{14}H_5O_2(NO_2)(OH)(ONa)$, and is easily soluble in water. The colouring-matter is not in such a fine state of division as those of the alizarin pastes, and if allowed to stand, it settles to the bottom.

In order to detect impurities or unchanged alizarin, a sample is dissolved in dilute caustic soda, filtered (if necessary), and treated with an excess of concentrated caustic soda. The nitro-alizarin is thereby completely precipitated as the sodium salt, and the filtrate can easily be tested for alizarin.

With alumina mordants, nitro-alizarin yields an orange; with iron mordants, a red shade of violet.

Application.—Alizarin-orange is used chiefly in calico-printing as a steam-colour, and is fixed just like alizarin. The colour for printing contains besides alizarin-orange aluminium nitrate, calcium acetate, and a thickener. It does not keep so well as that prepared with alizarin, since nitro-alizarin is a much stronger acid, and has a great tendency to form lakes, even in the cold. This may be checked to some extent by the addition of some acetic acid; nevertheless, the colour must be used as soon as prepared, otherwise it becomes red, and does not yield an orange, but a dull yellow.

This rapid change is a great drawback in the application of alizarin-orange.

According to Kielmeyer, the addition of calcium hyposulphite in place of calcium acetate will make the colour keep for at least a day. The introduction of aluminium sulphocyanide in place of aluminium nitrate will no doubt further the application of alizarin-orange. The colours prepared with this mordant are said to keep very well.

The material is invariably oiled before printing. After steaming, the colour is dull, and is only developed into a bright shade after treatment with a boiling soap solution.

If red prussiate of potash is added to the colour before printing, some Prussian-blue is formed along with the nitro-alizarin alumina-lake. Fine shades of *brown* can be produced in this manner.

Alizarin-orange is exceedingly fast; it is not attacked by solutions of chloride of lime, but it is destroyed by the joint action of chloride of lime and acids.

Alizarin-orange is used to some extent in wool-dyeing on account of the fastness of the shades produced. With an alumina mordant it yields bright shades of orange; with bichromate of potash, brown shades of orange. It is generally applied on wool along with other colouring-matters for the production of mixed shades. Thus when used along with coeruleïn, alizarin-blue, or alizarin, very beautiful effects can be obtained.

Detection on the fibre.—The colour is brightened by boiling with soap, but it is stripped with an orange colour by ammonia or caustic soda. If boiled with barium hydrate the sample is turned violet. An acid solution of stannous chloride strips with a yellow colour.

Alizarin-blue.
(Anthracene-blue.)

Alizarin-blue is prepared in the following manner:—
One part of dry and finely pulverised nitro-alizarin is

heated, with 5 parts of dehydrated glycerin, and 5 parts concentrated sulphuric acid, to 150°. After the reaction is over, the melt is boiled with an excess of water, when the colouring-matter passes into solution in the shape of its sulphuric acid compound. On cooling it falls in the form of a brown flocculent precipitate, which when washed with water loses its acid and becomes blue.

Alizarin-blue occurs in commerce as a dark-violet paste containing 10 per cent. of dry substance. The pure colouring-matter can be obtained from the dried paste by crystallising repeatedly from glacial acetic acid and naphtha. In the pure state it forms brown-violet needles, which melt at 270°, and may be partially sublimated. It is insoluble in water, and only sparingly soluble in benzene and alcohol.

The empirical formula of alizarin-blue is $C_{17}H_9NO_4$. It is closely allied to quinoline, C_9H_7N, which is obtained in a similar manner by heating nitro-benzene and glycerin with concentrated sulphuric acid.

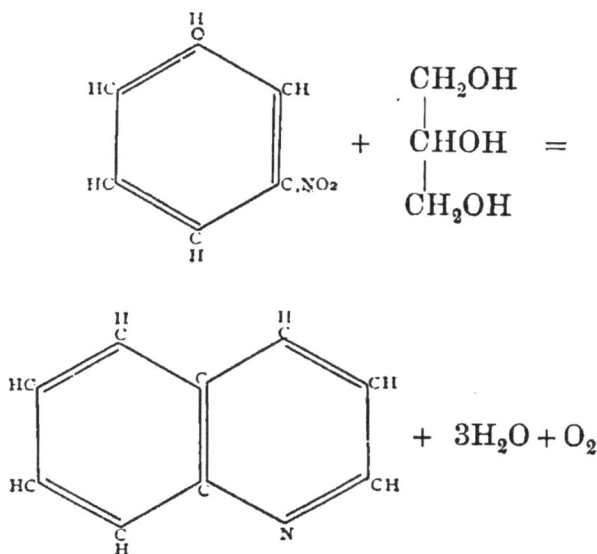

Alizarin-blue is formed according to the equation—

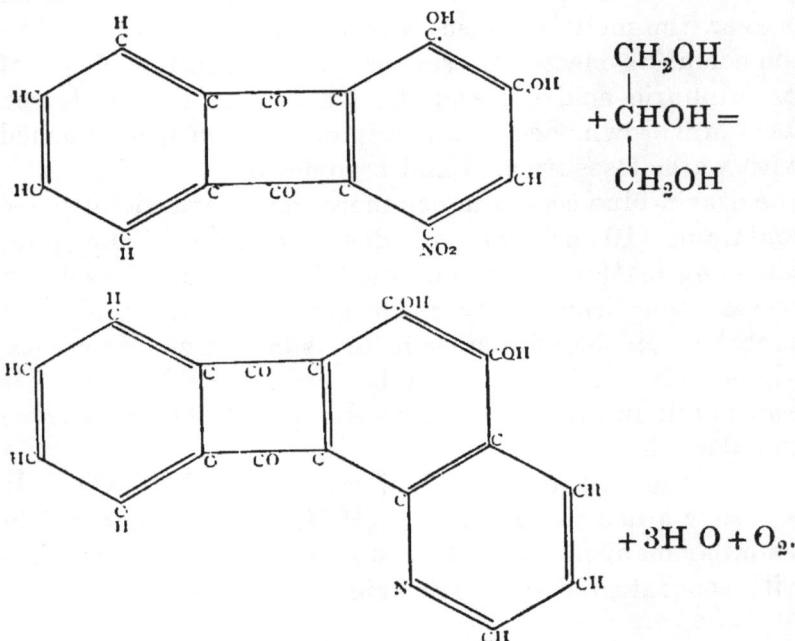

$$+ \begin{matrix} CH_2OH \\ | \\ CHOH \\ | \\ CH_2OH \end{matrix} =$$

$+ 3H\,O + O_2.$

Alizarin-blue is, therefore, the quinoline of alizarin. It contains both hydroxyl groups of alizarin unchanged, and is therefore an acid. On the other hand, however, the quinoline complex imparts basic properties to the compound, and alizarin-blue will therefore combine with acids to form unstable compounds. Thus it will crystallise from glacial acetic acid, and from dilute sulphuric acid as a brown salt containing one equivalent of acid. But these compounds are so loose that they are decomposed by washing with water. In concentrated sulphuric acid alizarin-blue dissolves with a red colour, while in arsenic acid and phosphoric acid it dissolves with a red-yellow colour.

With dilute caustic alkalies it yields a green-blue solution, from which it is precipitated by excess of alkali.

All the other salts are insoluble in water. They are obtained by precipitating the alkaline solutions of alizarin-blue with the corresponding metallic salts. Alizarin-blue yields with lime, baryta, and ferric oxide, greenish-blue, with nickel, blue, and with alumina and chromium, bluish-violet lakes.

The affinity of alizarin-blue for some bases is so great that in some cases it will expel even sulphuric acid from its compounds (e.g., from copper sulphate).

In alkaline solution it may be reduced with zinc powder to a red liquid, which, when exposed to the air, resumes its original colour, and therefore forms a vat like indigo.

Application.—The application of alizarin-blue for dyeing on the large scale is somewhat restricted. Its insolubility in water and its tendency to form insoluble lime-lakes are not in its favour. In vat-dyeing it cannot compete with indigo, on account of its high price.

For printing similar methods may be employed as those used for indigo. The colour for printing can be made up with stannous oxide and caustic soda, which reduces, and at the same time dissolves, the colouring matter. After printing, the colour is developed by exposure to the air and washing in running water. The results obtained by this method have, however, not been very satisfactory.

The best *steam colour* is obtained with chromium acetate. An addition of magnesium chloride or calcium chloride produces a faster, and at the same time a purer, blue.

The greatest drawback in alizarin-blue fixed in this manner is that it is not so fast to light as indigo.

If the colours printed or dyed on calico are exposed to sunlight, they become yellowish, and, after soaping, pale violet.

Alizarin-blue stands the action of oxidising agents better than indigo; it is not so easily attacked by chloride of lime, chromic acid, and alkaline ferricyanide of potash.

Alizarin-blue S.—Like coeruleïn, alizarin-blue can be transformed into a dry powder soluble in water. For this purpose, the commercial paste is mixed with a concentrated solution of sodium bisulphite, and after standing for 8–14 days, the liquid is filtered, and the solid compound is isolated by evaporating down or salting out.

Alizarin-blue S is a dark-purple powder, which is easily soluble in water. The brown-red solution is decomposed by strong acids and by soda; it is also decomposed if heated above 70°. The colour can, however, be mixed without undergoing any change with salts of chromium, lime, or magnesia, as well as with acetic or tartaric acid. The decomposition only takes place on steaming.

Alizarin-blue S is not only distinguished from ordinary alizarin-blue by its solubility and easier application, but also by its greater fastness to light. It has consequently replaced ordinary alizarin-blue to a great extent in printing and dyeing.

In wool-dyeing, alizarin-blue S promises to be a great success as a substitute for indigo. The wool is first mordanted with bichromate of potash and tartar, and afterwards dyed in alizarin-blue S. In the dyeing it is necessary to observe certain precautions. In the first place, the water should contain no lime, or, if lime is present, it should be neutralised with acetic acid. Secondly, it is necessary, in order to obtain full and even shades, to dye very gradually, working first for some time in the cold, and then gradually raising the temperature to a boil. In case the water contains bicarbonate of lime in solution, it should be neutralised before dyeing with acetic acid. Beautiful deep-blue shades can be obtained in this manner, which are as fast as indigo to light and milling, and, besides, possess a characteristic bloom.

Alizarin-blue S W is a cheaper form of soluble alizarin-blue, consisting simply of a strong solution of the bisulphite compound. It has replaced the powder to a great extent.

Detection on the fibre.—Alizarin-blue is turned bluish-green by alkalies, while dilute hydrochloric acid turns it violet. Nitric acid decolourises, or, in case the blue is dyed on wool with a mordant of bichromate, it produces an orange spot. An acid solution of stannous chloride turns the colour to a brownish-yellow. Solutions of soap, soda, and chloride of lime are without action.

Phosphoric acid (sp. gr. 1·435) gives an orange solution, which, when diluted with water, is turned blue again by ammonia.

The spectroscope may also be used for detecting alizarin-blue.

Alizarin-green S.

This colouring-matter is formed in the following manner:—By treating alizarin-blue with a large excess of fuming sulphuric acid, a product is obtained which by the action of alkalies or acids is converted into "alizarin-blue-green." If this latter product is heated with concentrated sulphuric acid to 120–130°, alizarin-green is formed, which after separating and washing with water is obtained in the form of fine bluish-grey needles. The commercial product forms a red-brown liquid, which smells strongly of sulphurous acid, and when boiled yields a blue precipitate. On wool mordanted with bichromate of potash it produces greenish shades of blue, resembling in appearance dull shades of indigo extract. The colour stands milling.

Detection on the fibre.—Boiling with dilute sulphuric acid removes the greater part of the colour. Stannite of soda turns the colour to a dull violet; the original colour is restored by washing and exposure to the air. Nitric acid produces a yellow spot with a violet border.

Alizarin-black S.

This colouring-matter is the sodium bisulphite compound of naphthazarin or dioxynaphthoquinon. Its composition is represented by the formula $C_{10}H_4O_2(OH)_2 +$ NaH SO_3.

Naphthazarin is formed by the action of zinc and strong sulphuric acid on dinitronaphthalene.

The commercial product forms a black-coloured paste, which dissolves in boiling water with a reddish-brown colour, which is turned to a fine blue on the addition of caustic soda.

Application.—On wool mordanted with bichromate of potash alizarin-black produces reddish shades of black. Dead blacks can be obtained by the addition of a little coeruleïn. With small percentages of the colouring matter, various shades of slate are obtainable, which are considerably faster to light than logwood shades. On calico, alizarin-black may be printed with a chrome mordant.

Galloflavin.

Galloflavin is formed by passing a limited quantity of air through an alkaline solution of gallic acid in water or alcohol at a low temperature. The empirical formula, $C_{13}H_6O_9$, is assigned to the new dyestuff produced in this manner; the constitution is not known.

The commercial product forms a dull-yellow paste, which dissolves in caustic soda with a brown-orange colour.

On wool mordanted with bichromate of potash, galloflavin yields shades resembling those obtained with fustic. It is also employed with a chrome mordant in calico printing.

The shades obtained with galloflavin on wool are not very fast to light.

ADDENDUM.

Tartrazine.

This colouring-matter is the sodium salt of disulpho-diphenylizine-dioxytartaric acid, the constitution of which is represented by the formula—

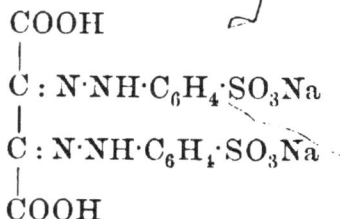

$$
\begin{array}{l}
COOH \\
| \\
C : N \cdot NH \cdot C_6H_4 \cdot SO_3Na \\
| \\
C : N \cdot NH \cdot C_6H_4 \cdot SO_3Na \\
| \\
COOH
\end{array}
$$

It is formed by acting (in aqueous solution) with phenylhydrazine monosulphonic acid on dioxytartaric acid.

The commercial product forms an orange-yellow powder, easily soluble in water, with an orange-yellow colour. In concentrated sulphuric acid it dissolves with a yellow colour. Tartrazine dyes wool yellow in an acid bath. The colour produced is fast to light, and also stands soaping fairly well.

INDEX.

A.

Absorption spectra, 13, 18.
  of mixtures, 22.
Acetanilide, 96.
Acid brown G, 262.
  colouring-matters, 31.
  green, 146, 147.
    S.O.F., 147.
  magenta, 160.
  violet 6 B, 175.
  yellow, 255, 266.
    R, 257.
    D, 258, 267.
Acridin, 106.
Adjective dyeing, 48.
Albumen, 56.
Aldehyde green, 185.
Alizarin, 294.
  absorption spectra of, 16.
  black S, 324.
  blue, 318.
    S, 322.
    shade, 299.
  carmine, 314.
  G, 299.
  green S, 323.
  manufacture of commercial, 291.
  orange, 316.
  pink, 302.
  red, 302.
    S, 314.
    V, 299.
  violet, 308.
  W.S., 314.
  yellow shade, 299.

Alkali-blue, 166.
  green, 148.
Alpha-napthol, 123.
  sulphonic acids of, 125.
Amaranth, 156, 268.
Amido-azo dyes, 251.
  sulphonic acids, 112.
Amines, 31.
  formation of, 94.
Aniline, 94.
  black, 201.
  blue, 161.
  dyes, 140.
  orange, 284.
  red, 148.
  salt, 96.
  yellow, 252.
Anisol, 118.
Azodiphenyl-blue, 187.
Anthracene, 86.
  brown, 315.
  blue, 318.
  oil, 72.
Anthraflavic acid, 299.
Anthragallol, 315.
Anthrapurpurin, 296.
Anthraquinon, 133.
  sulphonic acids of, 135.
Archil red A, 265.
Aromatic acids, 135.
  aldehydes, 129.
Artificial indigo, 236.
Auramine, 181.
Aurantia, 223.
Aureosin J, 235.
Aurin, 227.
Azaleïne, 156.

327

Azarin, 277.
Azo-black, 269.
   blue, 280.
   compouuds, 108.
   dyes, quantitative determina-
      tion of, 65.
   violet, 281.

B.

Bases in coal-tar, 70.
Basic colouring-matters, 30.
Basle-blue, 195.
Bavarian-blue, D.S.F., 167.
Benzal chloride, 87.
   green, 144.
Benzaldehyde, 130.
Benzene, 80.
   sulphouic acids of, 111.
Benzidine, 100.
   sulphouic acid of, 114.
   azo dyes derived from, 277.
Benzoic acid, 136.
Benzophenon, 132.
Benzopurpurin, 280.
      4 B, 279.
      6 B, 279.
Benzotrichloride, 88.
Benzoyl-green, 144.
   chloride, 87.
Benzylrosaniline-violet, 173.
Beta-napthol, 123.
   sulphonic acids, 126.
Biebrich-scarlet, 269.
Bismarck brown, 254.
Bisulphite compounds of the azo
   dyes, 276.
Bleu fluorescent, 225.
   de nuit, 161.
   lumière, 161.
Blackley-blue, 188.
Blue-black B, 269.
Boiled off liquor, 42.
Brilliant cotton blue, 168.
   Congo G., 279.
   croceïn, 265.
   green, 146.
   orange, 262.
   scarlet, 268.
   yellow, 222, 282.

C.

Canelle, 254.
Carbolic acid, 118.
Cardinal, 253, 156.
Carmin-naphtha, 265.
Carmoisin, 267.
Carnotine, 283.
Casein, 43.
Cellulose, 43.
Cerise, 156.
Chestnut-brown, 184.
China-blue, 168.
Chrome-violet, 196.
Chromophorous groups, 29.
Chrysamin, 279.
   R, 280.
Chrysaniline, 284.
Chrysaureïn, 266.
Chryseolin, 266.
Chrysoïdine, 252.
Chrysolin, 234.
Cinnamic acid, 286.
   aldehyde, 130.
Cinnamon-brown, 254.
Cloth-red B, 265.
      G, 265.
      S, 268.
Coal-tar colours, history of. 2.
   quantitative determination
      of, 64.
   classification of, 139.
Coal-tar, formation of, 69.
   distillation of, 71.
   constituents of, 74.
   raw products of, 75.
Coccinin B, 264.
Cochineal red A, 268.
Coeruleïn, 242.
Colorimetry, 64.
Colouring-matters, general chemi-
   cal properties of the, 27.
   optical properties of the, 5.
   methods of dissolving, 35.
   marks used for distinguish-
      ing, 7.
   as mordants, 56.
   testing of, 60.
   impurities in, 66.
Coloured inks, 36.

Coloured varnishes, 38.
Comparative dye-trials, 60.
Complementary colours, 6.
Corallin, 229.
Congo-red, 278.
Constitutional formulæ, 28.
Cotton, 43.
    blue, 168.
        R. 197.
Coupier's blue, 186.
Cresorcin, 121.
Cressols, 121.
Cressol red, 265.
Croceïn 3 B X. 267.
    orange, 262.
    scarlet, 3 B, 268.
        7 B, 269.
        6 R, 265.
Crystal-violet. 174.
    quantitative determination of, 66.
Cumidines, 99.
Cumidine-red, 264.
    scarlet, 264.
Cyanosin, 237.

D.

Dahlia, 170.
Dark green, 226.
Deltapurpurin 7 B, 279.
Diamido-benzenes, 96.
Dianthine B, 235.
Diazo compounds, 106, 261.
Dichroistic solutions, 17.
Diethylaniline, 104.
Dimethylaniline, 103.
Dinitrobenzenes, 90.
Dinitrotoluenes, 91.
Dioxybenzenes, 119.
Dioxytoluenes, 122.
Diphenyl, 83.
Diphenylamine, 102.
    blue, 162.
    orange, 258, 266.
Discontinuous spectra, 12.
Distillation of coal-tar, 71.
Dyeing, general principles of, 45–57.

E.

Empirical formulæ, 75.
Eosins, 234.
Eosin, absorption spectrum of, 14.
    A, 234.
    A extra, 234.
    B, 234.
    B. B., 236.
    B. N., 235.
    G.G.F., 234.
    5 G, 235.
    J, 235.
    orange, 235.
    soluble in spirit, 236.
    yellowish, 234.
Erythrosin, 235.
    B, 237.
Essence de Mirbane, 90.
Ethyl-blue, 163.
    eosin, 236.
    green, 146.
    violet, 174.
Ethylene-blue, 212.
Eurhodines, 196.

F.

Fat colours, 39.
Fast and loose colours, 57,
Fast-green, 144.
    blue R, 188.
        B, 188.
        2 B, 196.
        R, for cotton, 196.
    red A, 267.
        B, 265.
        C, 267.
        D, 268.
        E, 268.
    scarlet B, 269.
    yellow, 255.
        R, 257.
Fibres, chemistry of the, 39.
Flavaniline, 283.
    S, 284.
Flavopurpurin, 296.
Fluoresceïn, 232.
Fluorescence, 25.
Fluorescent resorcin-blue, 225.

Fuchsin, 156.
V, 156.

G.

Galleïn, 242.
Gallic acid, 137.
Gallocyanin, 198.
Galloflavin, 324.
Gallotannic acid, 137.
Gambine, 226.
Gentian-blue 6 B, 161.
Giroflé, 195.
Gold-orange, 257, 266.
Golden yellow, 266.
Green grease, 72.
Guernsey-blue, 188.
Guinea-green, 147.

H.

Heat, action of—on dyed fabrics,
58.
Heavy oil, 72.
Helianthin, 257, 266.
Heliochysin, 222.
Helvetia-green, 146.
Hessian-purple B, 282.
D, 282.
N, 281.
P, 282.
Hessian-violet, 282.
Hoffman's violet, 170.
Homogeneous light, 9.
Huile Lournante, 305.
Hydrazines, 108.
Hydrocarbons in Coal-tar, 69.
Hydrocarbons, the, 80.
Hydroquinon, 120.
Hydroxyl group, 33.

I.

Indigen, 186.
Indigo substitute, 188.
Indophenol, 199.
white, 200.
Induline, 186.
Imperial scarlet, 269.
Impurities, estimation of—in
colouring-matters, 6.
Iso-butylphenol, 122.
Isomeric compounds, 79.
Isopurpuric acid, 215.

K.

Ketons, 132.

L.

Lakes, 34.
Lauth's violet, 212.
Leuco bases, 142.
Light, action of—on dyed fabrics,
59.
Light, decomposition of—by the
prism, 7.
Light-green, 147.
Light oil, 72.
Liquid toluidine, 99.
Line spectra. 12.
Lutécienne, 236.

M.

Magdala-red, 193.
Magenta, absorption spectrum of,
14.
Magenta, 148.
manufacture of, 153.
detection of, 158.
S, 193.
Malachite-green, 144.
soluble in spirit, 145.
Manchester-yellow, 219.
Mandarin, 266.
G R, 267.
Maroon, 184.
Martius-yellow, 219,
Mauveïn, 196.
Metaxylene, 83.
toluidine, 98.
Metallic sulphides as mordants,
51.
Metanil-yellow, 267.
Meldola's blue, 197.
Methylene-blue, 208.
Methyl-blue, 163.
B I, 168.
Methyl water-blue, 168.
violet, 170.
6 B, 173.
green, 175.
eosin, 236.
orange, 257, 266.

Methyl-violet, absorption spectrum of, 14.
Methyldiphenylamine, 104.
Mesitylene, 83.
Mikado-orange, 282.
Milling, 58.
    action of, on dyed fabrics, 58.
Monoethylanaline, 102.
Monomethylaniline, 102.
Muscarin, 197.

N.

Naphthameïn, 207.
Naphthionic acid, 114.
Naphthols, 123.
    sulphonic acids, 124.
Naphthol-yellow, 219.
    S, 221.
    quantitative estimation of, 65.
Naphthol-green, 227.
    (alpha) orange, 266.
    (beta) orange, 266.
    orange, 266.
    black, 269.
Naphthylamines, 100.
Naphthylamine, sulphonic acids of, 114.
Naphthalene, 84.
    tetra chloride, 88.
Naphthylene-blue R,
    red, 193.
    violet, 207.
Neutral colouring-matters, 35.
    oils, 81.
    blue, 195.
    violet, 196.
    red, 196.
New coccin, 268.
    R, 265.
New green, 144.
    red, 156.
    red L, 269.
    blue, 196.
    yellow, 259, 267.
        L, 255.
Nicholson's blue, 166.
Night-blue, 184.
    application of, for quantitative determination of acid colours, 65.

Nigrosine, 186.
Nile-blue, 197.
Nitro-alizarin (β), 316.
Nitro-sulphonic acids, 112.
    benzene, 89,
    toluenes, 91.
    xylenes, 92.
    naphthalenes, 92.
Nitroso-dimethylaniline, 104.

O.

Oil mordants, 51.
Opal-blue, 161.
Orange, 266.
    I., 266.
    II., 266.
    III., 257, 266.
    IV., 258, 267.
    G, 262.
    G S, 258.
    No. 3, 262.
    G T, 263, 267.
    M, 258, 267.
    M N, 267.
    T, 267.
    R, 267.
Orceïn, 122.
Orcin, 122.
Orchil substitute, Poirrier's, 262.
Orthoxylene, 83.
Orthotoluidine, 98.
Orthonitrophenyl-propiolic acid, 286.
Oxyanthraquinon, 299.
Oxyazo dyes, 259.
    compounds, 108.
Oxytoluenes, 121.

P.

Para-rosaniline, constitution and synthesis of, 149.
Paratoluidine, 98.
Paraxylene, 83.
Paris violet, 170.
Perspiration, action of, on dyed fabrics, 58.
Phenol, 118.
Phenols in Coal-tar, 70.
    formation and behaviour of, 116.

Phenicienne, 218.
Phenol phthaleïn, 231.
Phenosafranine, 189.
Phenyl-brown, 218.
  violet, 169.
Phenylene-brown, 254.
  diamine, 97.
Philadelphia-yellow, 284.
Phloroglucin, 121.
Phloxin, 237.
  P, 237.
  T A, 237.
Phosphine, 284.
Phthaleïns, 230.
Phthalic acid, 138.
Picoline, 105.
Picric acid, 213.
  absorption spectra of, 14.
  quantitative determination
    of, 65.
Pink, 190.
Pitch, 72.
Polychromine, 283.
Primary colours, 6.
  amines, 31, 94.
Primrose, 236.
Primuline, 282.
Propiolic acid, 286.
Prune, 198.
Pseudocumene, 83.
Purpurin, 297.
Pyrocatechin, 119.
Pyrogallic acid, 120.
Pyridine, 105.
Pyrosin, 235.

Q.

Quinaldine, 106.
Quinoline, 105.
  sulphonic acid, 115.
  blue, 245.
  red, 246.
  yellow, 245.

R.

Rancid Gallipoli, 305.
Red corallin, 229.
Red-violet 4 R S, 175.
  5 R extra, 170.

Regina purple, 173.
Resorcin, 119.
  brown, 264.
  green, 226.
  yellow, 266.
Rhodamine, 241.
Roccellin, 267.
Rosaniline, 149.
  group, 140.
  salts, properties of the, 155.
Rosazurin B, 279.
Rose B, soluble in water, 235.
  J B soluble in spirit, 236.
Rose bengale, 237.
Roseïn, 156.
Rosolan, 196.
Rosolic acid, 227.
Rosonapthylamine, 193.
Rosophenolin, 230.
Rubidin, 267.
Rubine, 156.
  S, 156.
Rubeosin, 236.

S.

Safranine, 190.
  B extra, 189.
Safrosin, 235.
Salicylic acid, 136.
Salt-forming groups, 30.
Saturated colours, 6.
Scarlet G, 263.
  2 G, 262.
  4 G B, 262.
  G R, 263.
  G T, 263.
  R, 263.
  2 R, 264.
  3 R, 264.
  6 R, 268.
  3 R B, 269.
  4 R B, 268.
  6 R B, 269.
  R T, 263.
Scroop silk, 42.
Secondary amines, 31, 101.
Secondary azo compounds, 261.
Shades, 6.
Silica as a mordant, 53.

Silk, 40.
Soap as a mordant, 51.
Solid-green, 144.
    J, 146.
    violet, 198.
    yellow, 261.
Soluble-blue, 168.
    eosin, 234.
    primrose, 235.
Solvents for colouring-matters, 36.
Souple silk, 42.
Spectra, 11.
Spectroscope, 10.
Spectrum, 9.
Spirit-blue O, 161.
    soluble-blue, 161.
    primrose, 236.
    violet. 169.
Stilbene, 83.
    azo dyes derived from, 277.
    colours derived from, 281.
Substantive dyeing, 45.
Sulphanilic acid, 113.
Sulphine, 283.
Sulpho group, 33.
Sulphonic acids, 34, 109.
Sulphur as a mordant, 50.
Sun-gold, 222.

T.

Tannin mordants, 54.
Tartar emetic, 54.
Tartrazin, 325.
Tetrabromfluorescein, 234.
Tetramethyldiamidobenzophenon, 132, 133.
Tetraiodfluorescein, 235.
Tetrazo compounds, 261.
Tertiary amines, 31, 102.
Thionin, 212.
Tintometer, The, 64.
Tolidine, 101.
Tolidines, azo dyes derived from, 277.
Toluene, 81.

Toluidines, 97.
Toluidine, sulphonic acids of, 113.
Tolylenediamines, 99.
Trioxybenzenes, 120.
Triphenylmethane, 149.
Tropæolin D, 257, 266.
    O, 266.
    O O, 253, 267.
    O O O No. 1, 266.
    O O O No. 2, 266.
    R, 266.
Turkey-red oil, 52.

U.

Uranin, 234.

V.

Vesuvine, 254.
Victoria-blue, 183.
    green, 144.
    yellow, 219.
Violaniline, 186.
Violet 5 R, 170.
    6 B, 174.
Viridine, 148.

W.

Water-blue, 168.
Wool. 42.
    scarlet R, 263.

X.

Xanthoproteïc acid, 43.
Xylenes, 81.
Xylenols, 122.
Xylidine, 99.
    sulphonic acids, 113.
    red, 264.
    scarlet, 264.

Y.

Yellow corallin, 229.
    T, 266.
    W, 257.

Butler & Tanner, The Selwood Printing Works, Frome, and London.

www.ingramcontent.com/pod-product-compliance
Lightning Source LLC
Chambersburg PA
CBHW021457210326
41599CB00012B/1034